DAVID O. MCKAY LIBRARY
3 1404 00753 7688

Impure Cultures

D0744932

JAN 29 2004

WITHDRAWN
JAN 31 2023
DAVID O. McKAY LIBRARY
BYU-IDAHO

PROPERTY OF:
DAVID O. McKAY LIBRARY
BYU-IDAHO
REXBURG ID 83460-0405

SCIENCE AND TECHNOLOGY IN SOCIETY

Series Editors

Daniel Lee Kleinman
Jo Handelsman

Impure Cultures

University Biology and the World of Commerce

Daniel Lee Kleinman

The University of Wisconsin Press

The University of Wisconsin Press
1930 Monroe Street
Madison, Wisconsin 53711

www.wisc.edu/wisconsinpress/

3 Henrietta Street
London WC2E 8LU, England

Copyright © 2003
The Board of Regents of the University of Wisconsin System
All rights reserved

5 4 3 2 1

Printed in the United States of America

Library of Congress Cataloging-in-Publication Data
Kleinman, Daniel Lee.
Impure cultures: university biology and the world of commerce / Daniel Lee Kleinman.
p. cm.—(Science and technology in society)
Includes bibliographical references and index.
ISBN 0-299-19234-2 (pbk: alk. paper)
1. Biology—Research—Economic aspects. 2. Education, Higher—Economic aspects. 3. Academic-industrial collaboration. I. Title. II. Series.
QH315 .K54 2003
570´.72—dc21 2003007233

To Jo Handelsman

and

The Gorillas in My Midst

Contents

Preface

I picked up the local newspaper some months back to find that the governor of our state had decided to move forward with the so-called BioStar Initiative—a massive public-private partnership to build state-of-the-art facilities for the biological sciences at the University of Wisconsin–Madison. First proposed by then-governor Tommy Thompson in January of 2000, the terms of this $317 million initiative require that the university administration convince corporate donors, foundations, and individuals to provide half of the financing for the effort. The aim of the project is to make the University of Wisconsin a star among academic institutions undertaking commercially relevant biological research and explicitly to "enhance" the state's biotechnology industry.

Such large scale collaborations between public and private institutions are increasingly common. In December of 2000, Governor Gray Davis of California awarded $300 million to three campuses of the University of California system to develop institutes that will serve as hothouses for commercially relevant high tech research, and the state money is to be matched 2-to-1 with funding obtained by university officials from industry, individual donors, or the federal government. As in Wisconsin, the aim of the project is to allow the state to profit from developments arising out of the new knowledge economy. In the California case, state tax dollars are largely directed toward bricks and mortar. Much of the operating money will come from high tech companies like Qualcomm and Sun Microsystems.

In the new knowledge economy, universities are central institutions. The high tech inventions that drove the economic dynamism that characterized the 1990s trace their beginnings to our institutions of higher education. In the biotechnology industry, for example, the underlying discoveries originated in the academy. The most central of these, of course, is the work done by Herbert Boyer at the University of California–San Francisco and Stanley Cohen at Stanford in the 1970s. Their research made recombinant DNA—and consequently the biotechnology revolution—possible. But much of the basic work on specific crops that provides the foundation for agricultural biotechnology was and continues to be done in university settings as well.

At the same time that universities are increasingly regarded as sources of national wealth, these institutions are becoming ever more commercialized. Yes—private industry funds some university research, but more dramatically, the ethos of commerce is thoroughly enveloping academic life. Faculty members are increasingly encouraged to patent their research results independent of whether the research was supported with industry funding. At my own university, the office of University-Industry Relations not long ago sponsored a workshop on "Business Planning for Entrepreneurial Scientists, Engineers and Technology Companies." Scholars are expected to become entrepreneurs. A proposal was recently advanced at the University of Wisconsin to create a masters degree program in biotechnology to serve the needs of the state's biotechnology industry for trained technicians. Every day the university is looking more like a high-tech trade school.

For nearly two decades now, critics of university-industry relations and policymakers have been troubled by the ways in which commercially motivated collaborations between university biologists and science-based companies can skew research agendas, prompt inappropriate restrictions on the flow of information, and create conflicts of interest. I do not believe such concerns should be dismissed, but I also do not think they adequately reflect the massive transformation of the university that is underway. Direct relationships between corporate patrons and academic scientists—the focus of much analysis and policy debate—constitute only a small part of what some analysts call "academic capitalism" (Slaughter and Leslie 1997).

It is my view that only by studying the thoroughly systemic effects of the world of commerce on the university can we truly comprehend what is happening to higher education in the United States. In this book, I explore one piece of this puzzle: the pervasive influence of the world of commerce on academic biology. I use what I learned from six months as a participant observer in one university biology laboratory as a springboard for exploring issues at the heart of the transformation of academic biology. I analyze several features of the subtle landscape where academic science and the commercial world are fused. While a great deal of research examines how corporate funding can directly shape research agendas, I instead investigate how corporate domination of a field of scientific investigation early in its development can indirectly affect the questions that are asked and the answers that are acceptable at a later time, even if the later research is not funded by industry.

In contrast to the focus of much debate over the past two decades on university-industry relations in the biological sciences, I do not consider how increased concern with intellectual property protection can lead to a culture of secrecy in the university, but rather how widely held assumptions about the nature of intellectual property protection affect the actions that academic scientists take, as well as the interactions between university scientists and corporations. I investigate how intellectual property protection may actually decrease, rather than increase, scientists' control over their work. I also examine what the frenzy over intellectual property protection means for the future of the "knowledge commons"—the world of accumulated ideas upon which all future innovation depends. I not only look at how the commercially developed tools of academic biology can advance research—a position commonly assumed, but also consider how these same tools may constrain scientific practice, as well as compromise the breadth and quality of graduate training.

This book can be read strictly as a study of the character of academic biology at the turn of a new millennium. But it is more than that. Throughout the book, and especially in chapters 2 and 6, I engage in a sustained debate with some of the key conceptual assumptions underlying the interdisciplinary field known as science studies. I suggest that the case of the commercialization of academic biology underscores the importance of an analysis of science that seriously considers the nature and influence of structure and power. Throughout the book I develop and utilize my understanding of structure and power in an attempt to understand and clarify the character of academic biology at the new millennium.

We stand at the threshold of a new knowledge intensive economy. Here, the boundary between the worlds of commerce and scholarship constantly shifts and blurs. The two realms are becoming fused. We can passively watch this transformation, or we can attempt to shape and direct it. I hope that by providing a new perspective on this inadequately considered topic, this book can prompt more urgent concern among scholars, policymakers, and citizens, and encourage them to work actively to influence the changing character of the university in the new millennium.

Acknowledgments

This book was a number of years in the making, and I incurred many debts along the way. While I taught at Georgia Tech, I was fortunate to have several good friends with whom I could exchange ideas and share a beer. These include Jon Schneer, Joan Sokolovsky, Andrea Tone, and Steve Vallas. I am pleased that Steve and I continue to collaborate, and my contact with the others is ongoing as well. In addition to these Tech colleagues, I want to thank Bob McMath and Greg Nobles for all their help in securing the leaves during which I did the research for this project, and Denise Corum, who was always a supportive presence in the School of History, Technology, and Society at Georgia Tech.

In Wisconsin, I have benefited from the support and intellectual companionship of Fred Buttel, Chas Camic, Jess Gilbert, Abby Kinchy, and Jack Kloppenburg. Jack, along with Allen Hunter and Scott Frickel, read an early draft of the manuscript. All three of them provided me with vital comments and suggestions, and I continue to benefit from an ongoing conversation with Scott on the many issues taken up in this book. Later in the process, Brian Martin and David Hess read the manuscript as peer reviewers for the University of Wisconsin Press. Their comments allowed me to tighten the book substantially. I am also pleased that the review process helped initiate my contact with these scholars.

I wrote a good deal of this book during two stays in England, where I was affiliated with Clare Hall, University of Cambridge.

Members of the staff and fellows at Clare Hall made my time there especially productive. While in Cambridge the first time, I met and became friends with Paul Morrish and Rebecca Stott. I am grateful for their friendships.

Beyond my assorted academic homes, I received support of various kinds from several people. I benefited from discussions about ethnography with Stacy Wolf, and conversations about science and technology studies with Kelly Moore. Larry Cohen read the book's first chapter and provided me with helpful suggestions for improvement. My mother, Barbara Kleinman, is my informal clipping service. She regularly turns up news articles relevant to my work; several of her finds were particularly useful in this project. My father, Gerald Kleinman, has also been a help, providing me with both legal advice and typographical oversight. Finally, there is no way I can adequately thank my partner, Susan Bernstein, and daughter, Flora Berklein, for tolerating me at my most difficult and for every day providing the kind of crucial sustenance that is too easy to overlook.

This project would not have been possible without financial support. I received an early research grant from the Georgia Tech Foundation, and a National Endowment for the Humanities Fellowship for University Teachers allowed me a block of time away from teaching to draft this book.

To move from manuscript to book requires a publisher, and I want especially to thank the director of the University of Wisconsin Press, Robert Mandel, who took an interest in my project and has made a broader commitment on the Press' behalf to science and technology studies. My thanks also are due to Tricia Brock, Erin Holman, Sharon van Sluijs, and others at the Press who helped prepare the manuscript for publication.

Finally, I owe my largest debt of gratitude to the scientists and others who were members of the Handelsman lab or affiliated with it when I did my field work. These include: Robert Goodman, Scott Bintrim, Mark Bittinger, Elizabeth Emmert, Lynn Jacobson, Laurie Luther, Jocelyn Milner, Eduardo Robleto, Eric Stabb, Sandy Stewart, and Liz Stohl. I want especially to thank Elizabeth Emmert and Eric Stabb, both of whom read this book in manuscript form and provided helpful comments, as well as Steve Stevenson who participated in and agreed to be interviewed on the bioregional biocontrol project (see chapter 6). Of course, this project would not have been possible without the unwavering support and regular guidance and assistance of Jo Handelsman. She opened her lab and professional life to my study and in the process brought science alive for me.

Impure Cultures

1

Impure Cultures

Every week, members of Professor Jo Handelsman's laboratory congregate in the "community room" for Chinese food. These gatherings are generally lighthearted. Anything and everything can provide the basis for humor. On a Thursday in July of 1995, lab members finished the entrees they ordered and turned their attention to the fortune cookies sitting on a crumpled white paper bag on the table. Some of the scientists read their fortunes aloud. Laughter abounded. An associate scientist in the lab cracked open her cookie and read: "You will prosper in the field of high technology."

This was not a prediction of the future as much as an accurate assessment of the present status of the Handelsman lab. One of several laboratories that make up the Department of Plant Pathology at the University of Wisconsin–Madison, the lab run by Jo Handelsman uses the latest techniques and concepts in the biological sciences in its efforts to fundamentally understand plant diseases and prevent them. The dominant—though not the sole—concern of the scientists Handelsman taught and managed in the mid-1990s was biocontrol, the use of microorganisms to suppress plant disease. Their focus of attention was a strain of *Bacillus cereus* they call UW85, a variety of bacteria that lab members discovered protects agricultural crops from two diseases, damping off and root rot, which are caused by a microorganism called *Phytophthora* (Gallepp 1994; Handelsman et al. 1990). In its research, this lab, which straddles the boundary of so-called basic and applied science,[1] has prospered by virtually every measure.[2] The Handelsman scientists' efforts in the biological-control area and the related work undertaken both in the lab and in collaboration with other labs has led to a slew of peer-reviewed scientific publications. A number of patents filed both in the United States and abroad bear the names of Handelsman lab workers. Although the procurement of funding in the academy is inevitably an anxiety-producing enterprise, Handelsman manages to keep her lab adequately supported through a wide range of governmental and industrial sources; her graduate students complete their research and find employment in a tight job market.

The Handelsman laboratory is one that utilizes the techniques flowing from the revolution in molecular biology that are now in widespread use in the life sciences. It is involved in simultaneously exploring fundamental biological questions and producing work that has practical potential. Thus, it is a laboratory that provides an excellent window into the practice of academic biological science at the turn of the century.

More specifically, my interest is in the ways that what might broadly be termed the *world of commerce* shapes the everyday practices of academic laboratory science. From this perspective, the Handelsman lab is a complicated case. The laboratory is a recipient of industry support, it patents its inventions, and it is committed to producing work that is useful beyond the walls of the academy. At the same time, Jo Handelsman is driven by intellectual curiosity and views her teaching role as preeminent. She will not let commercial concerns or commitments distract or divert her students from their education. She does not accept contract research. Industry gifts to her laboratory must have no restrictions placed on them. Regarding questions of intellectual property, her priority is always to facilitate the dissemination of research results to her scientist colleagues.

Despite the complicated nature of the lab, I joined Handelsman in the spring of 1995 as a participant-observer in my effort to understand the relationship between industry, intellectual property, and academic science. For six months, I took part in a wide range of laboratory activities. I attended weekly meetings in which ongoing laboratory research was discussed, and occasional meetings with representatives of industry and the university's patent agent. I was taught to screen soil samples from Costa Rica in search of *Bacillus cereus* strains that produce the novel antibiotic zwittermicin A (zmA), which lab workers initially found in *Bacillus cereus* from Wisconsin soils (Silo-Suh et al. 1994). Over the summer, I was responsible for organizing meetings to discuss the field research jointly undertaken by the Handelsman lab and the laboratory of Handelsman's colleague Robert Goodman. I joined lab workers in field harvests and plantings, and I occasionally helped out in the greenhouse. I continued for a year as a part-time observer of the lab after the end of my six month full-time stay.

My field data, which include many hours of interviews with Professor Handelsman and other members of the lab, as well as a battery of laboratory documents, provide the foundation for my analysis. These are supplemented with a wide range of documentary and secondary sources I explored, which enabled me to place my experience of life in the Handelsman laboratory in context, and to use the laboratory as a springboard to discuss this environment.

In general, this book can be seen as a sustained debate regarding two issues. First, throughout the book I examine how I believe we can productively think about commercial influences—or university-industry relations (UIRs), as they are known—on university biology. Here, my claim is that by placing the focus of our attention on possible threats to the university from direct and explicit relationships between university scientists and commercial concerns, we have neglected to notice the less overt, but far more pervasive effects of commercial factors—of commercial *culture*—on the practice of university biology. Second, I argue against several theoretical traditions in science and technology studies; I suggest that the analysis of organization or structure provides a better understanding of the practice of university science than an analysis that focuses on agency, especially where commercial matters are at stake. In the final chapter, I add to this critique the assertion that, contrary to much work in science and technology studies, there is explanatory value in giving conceptual priority to the "social" over the "technical."

In this chapter, I do two things. First, I introduce readers to the Handelsman lab. Here, I describe the nature of work in the lab, how laboratory scientists spend their time, and how life in the laboratory is organized. My discussion illustrates the fundamentally heterogeneous and boundary-spanning character of life and work in the Handelsman laboratory. The most important point of this description, however, is to illustrate that the lab's direct and explicit connection to the commercial world is a rather minor feature of laboratory life for Handelsman and her colleagues. If one seeks egregious violations of academic norms, one will not find them here. Thus, to understand the impact of the world of commerce on the life of the Handelsman lab, one must use a different perspective, a different lens.

In the second section of the chapter, I reflect on my experience as an ethnographer in the Handelsman lab and explore how the lens I used to study the lab may explain some of the conflicts that developed between lab workers and me. Finally, I conclude this chapter with an outline of the chapters to come.

The Handelsman Laboratory

In part as a consequence of Handelsman's commitment to scholarship and education, it would be possible to spend six months in her laboratory without seeing the kinds of practices that raise concern among analysts of university-industry relations in the biological sciences, practices such as corporate domination of university research agendas, research shrouded in secrecy, and conflicts of interest. I suspect the same would be true if one were to observe many university biology laboratories across the country. But what might a casual observer, an undergraduate student helper, or perhaps even a Ph.D. biology student see in the Handelsman lab? How might such a person describe what goes on?

Crammed on a portion of the sixth floor of Russell Laboratories at the University of Wisconsin–Madison is the laboratory complex of Professor Jo Handelsman. The lab includes four office spaces, a meeting room, and five laboratory work spaces depending on how one counts. The staff, including technicians, staff scientists, an administrative assistant, undergraduate students, graduate students, and postdocs varies from about ten to fifteen people. On any given day, a visitor is confronted with a cacophony of sound—pipette tips rhythmically hitting plastic containers, the din of a local pop-music station, complaints about significant others, arguments about university policy, discussions about the unreliability of the polymerase chain reaction (PCR). The walls tell visitors something about the institutional location of the lab: attached there are photos of fields of alfalfa and of lab workers beside a pick-up truck; posters announce seminars in microbiology, molecular biology, plant breeding. A clipped press account tells the story of the lab's research and the promise that lab inventions may hold for Wisconsin farmers.[3]

Now a full professor and a leader in her field, Jo Handelsman came to the Department of Plant Pathology in the University of Wisconsin's College of Agricultural and Life Sciences (UW CALS) as an assistant professor in 1985. The origins of Handelsman's field of plant pathology as a formal discipline are intimately linked to the land grant system of which UW CALS is a part. It was the 1862 passage of the Morrill Act in the United States that laid the groundwork for the land grant system and so of plant pathology (Rossiter 1979, 213; Axt 1952, 37; Kloppenburg 1988). With passage of the act, grants of land were made by the federal government to each state. States could build new agricultural and engineering-oriented universities on this land or sell the land and use the money to fund a new state institution. Land grant universities were intended to "counter the elitism of private universities" (Busch, Lacy, Burkhardt, and Lacy 1991, 92) and serve the interests and needs of America's rural population. According to Donald Stokes, the mission of land grant universities involved the melding of science and technology (1997, 39; see also Kloppenburg 1987; Buttel et al. 1986).

In the first years after passage of the Morrill Act, plant pathology was an underdeveloped field in the United States. In 1871, the U.S. commissioner of agriculture supported plant disease research by minimally trained scientists. When plant pathology was first taught in American universities it was part of botany courses (Stevenson 1959, 19). With the passage of the Hatch Act and the development of state experiment stations and agricultural colleges, plant pathology research and teaching became more prominent. It was not, however, until the early twentieth century that Cornell University established the first independent department of plant pathology (Rossiter 1979, 232; Whetzel 1918). But throughout its early history, a commitment to solving the plant disease problems facing farmers was central to the identity of plant pathology in the United States.

The Handelsman lab is part of this legacy. Working at a land grant university, Handelsman and her colleagues are committed to undertaking research of practical value. But like the research of the agricultural scientists in whose footsteps workers in the Handelsman lab follow, the research in Handelsman's lab cannot be accurately described as simply applied. Rossiter explains that nineteenth century agricultural scientists "dealt with practical problems that arose in certain economic contexts, but they were not really 'applying' well-established theoretical principles to practical problems. Instead, they were discovering new ideas and principles that were also highly useful" (1979, 240).

This concept echoes in descriptions of the kind of research to which scientists in the Handelsman lab are attracted. One researcher, then a Ph.D. candidate in the lab, told me that Handelsman's studies appealed to him "because you could do basic research and use a lot of the kinds of tools I like to use or think are kind of neat in a lab where you are really trying to get back to something in the real world—the environment: how the bacteria are doing something in their environment." Or as a Ph.D. student who worked in Robert Goodman's lab, but also on UW85, put it: "The basic is usually more interesting, but only if you can tie into something that will be applied or useful." An associate scientist in Handelsman's lab indicated that she was interested in using "basic or fundamental approaches" to answer applied questions, and another graduate student said she was attracted to the lab because it sits at the "cusp" of so-called basic and applied research. Still another Ph.D. candidate suggested that he was interested in "how things work"—his definition of basic research—with the aim of applying this understanding to "a new situation or application."[4]

A description of some of the work undertaken in the lab while I was there reinforces the sense that the research of the Handelsman lab straddles the fence that separates what are traditionally termed applied and basic research. Handelsman disparagingly refers to less rigorous approaches to microbial biocontrol as "spray and pray." In these cases, because workers have a very limited underlying knowledge of the biocontrol agents they use, they have no way to understand how and why a biocontrol agent works or how these microbes could be made more effective. By contrast, researchers in the Handelsman lab take a multifaceted approach to understanding how UW85 works, and think of the microorganism as part of a system that includes plants and other soil microbes. When I was in the lab, for example, one graduate student was doing research in an effort to explain how and why UW85 is inhibited by alfalfa seed exudate. What is the chemical produced by alfalfa seed and how does it work to inhibit UW85? Because alfalfa is an important crop in Wisconsin, and because much of the field testing of UW85 has been with alfalfa, understanding how and why alfalfa can limit the biocontrol capacity of UW85 is crucially important to maximizing the effectiveness of the disease suppression agent. At the same time, alfalfa and the soil in which it is planted are part of a complex system, and this scientist's project offers the possibility of understanding one dimension of that system.

Another project that was ongoing while I was in the lab involved an attempt to understand how the novel antibiotic produced by UW85—zwittermicin A—inhibits target organisms. The Ph.D. candidate in charge of the effort was working at the genetic level and attempting to ascertain which genes code for proteins that lead some organisms to be sensitive to zmA. Understanding the relationship between antibiotic producing microorganisms and other organisms within the same ecosystem offers important general insights into the world of microorganisms. In addition, by understanding how zmA works, researchers might be able to maximize the efficacy of UW85 in the field or engineer biocontrol agents that are more efficient than "naturally" produced organisms. Beyond biocontrol, understanding antibiotic resistance offers potential practical payoffs in medicine where human resistance to certain antibiotics is an increasingly significant problem (Brown and Boseley 1998a and 1998b).

While I was in the lab, one Ph.D. scientist was working at the other end of the zwittermicin problem. She was using genetic tools to understand how UW85 produces zwittermicin A. This involves locating the gene or genes that are responsible for zmA production. There is an effort here to understand the genetic functioning of a common soil bacteria, *Bacillus cereus,* but this work could also provide the basis for efficient ways to locate strains of *Bacillus cereus* that are superior zmA producers than UW85. It could also aid in the genetic manipulation of some *B. cereus* strain to create better antibiotic producers.

One other ongoing project in the Handelsman lab is worth mentioning. This is a project that was initiated by a postdoctoral fellow working with Robert Goodman and continued after the postdoc's departure by a lab technician. It involves an attempt to breed alfalfa varieties that enhance the efficacy of UW85. The basic biological consideration here is the interaction between plants and microorganisms and, in particular, a symbiotic relationship that could allow alfalfa plants to ward off disease. The practical benefit is clear.

Where might a lab visitor learn about this research? Perhaps at one of the many meetings held in the lab or in the plant pathology department. Weekly lab meetings are devoted to discussions of laboratory research in progress. These meetings typically begin with announcements on array of issues, including the availability of money for purchase of supplies, the need to take better care of equipment, and upcoming departmental or university-wide lectures or seminars. Sometimes Handelsman seeks the advice of lab members on such matters as whether to bring new graduate students into the lab. After this initial announcement period, the floor is turned over to a student or other laboratory scientist, who describes recent developments in her or his work.

These meetings appear to serve several functions. First, scientists-in-training are given experience in presenting their work. Lab members told me that they work hard in preparation for these talks. They want to do a good job before their colleagues and are relieved when they are finished. Second, researchers who are having trouble either practically (making experiments work) or conceptually (organizing their experiments) have an opportunity to gather advice. One week a technician described the difficulties she was having growing the zoospores she needed to conduct her experiment; participants in the meeting provided a myriad of possible explanations for her problem and many suggestions for solving it.

Third, students may be given advice by Handelsman and other senior members of the lab on how to construct a compelling narrative about their research. After one of the people in the lab finished his talk, for example, Handelsman suggested to him that he "need[ed] to justify the genetics . . . because it is not the approach a biochemist would take." He needed to make it clear, furthermore, that he was asking "important questions" and not just trying to produce specific results. As a veteran in the field,

Handelsman is aware of how the lab's research relates to other areas it touches and how diverse audiences might respond to it. Consequently, a great deal of the advice she gives concerns how to frame and present research. Other suggestions she gave in this case revealed her orientation to science. She recommended he avoid using the passive voice in his presentation. Not only would this make the talk more interesting to a listener, but it is a "political issue," she said. Science does not just "get done." People do science.[5]

Finally, these meetings provide a chance for Handelsman to catch up on work underway in her lab and to contribute to its further development. She might recommend additional experiments or suggest a research literature to explore. An interaction between Robert Goodman and his postdoc at a joint lab meeting illustrates how a lab leader might contribute to the work of a junior scientist in the lab. When the postdoc exploring the interaction between alfalfa and microorganisms gave a talk, for example, Goodman looked at the data he presented and suggested that an application that combined UW85 with a commonly used fungicide might be advantageous for economic and environmental reasons. This is something that could be used to promote the commercial potential of UW85, and it is something that Goodman apparently had not been aware of before.

In addition to these lab meetings, every week students and Handelsman and Goodman crowd into the community room for "journal club." Members of the labs each monitor one or more journals that cover issues of interest to the Handelsman and Goodman labs. The sharing of responsibility for keeping up to date in the many fields that impinge upon the work done on the sixth floor of the Russell Laboratories permits wider coverage, and the requirement that students be responsible for some of this material ensures that they will keep abreast of new developments. At these meetings, lab members discuss the papers they have read, outlining the experiments that were undertaken and results obtained. Often, although not always, the presenter or another club participant evaluates the article under discussion. Handelsman frequently asks whose lab the research comes from so she can provide context for the students.

In addition to lab meetings, there are departmental gatherings. There are formal departmental seminars that take place in a large lecture hall. These are often presented by students finishing a degree or by visiting scientists. And then there is "Friday at Four," a weekly meeting where members of various labs in the plant pathology department take turns presenting reports on some aspect of their work. Often the laboratory leader introduces the discussion, and this is followed by a presentation from one of the lab members. Discussion is informal, over beer, juice, and pretzels, with questions ranging from the most basic to the more probing, including queries that call into question a procedure or approach.

In the summer, departmental talks subside, and while I was in the lab an additional meeting was added for a number of students in the Handelsman and Goodman labs. "Writing club" was an opportunity for students to read and critique each other's work-in-progress and to receive responses from Handelsman and Goodman. Handelsman and Goodman help students frame their research, interrogate their assumptions, and direct them to relevant literatures.

An observer would likely note that during the summer when Handelsman is not formally teaching, when grad students are not serving as teaching assistants or taking classes, laboratory life changes. Because the laboratory is concerned with microbial ecology and plant-microbe interactions in agricultural fields, field research and the attendant planning and organization dominate the lives of many people in the lab. During the summer I was in the lab, four studies were undertaken. One involved soybeans and investigated antibiotic production, plant health, and microbial ecology. A second project studied multiple strains of *Bacillus cereus* and different alfalfa cultivars at four field sites. A third study was an investigation of transgenic potatoes and their effects on microbial communities. The fourth project involved tomatoes and the increase of seed production.

Practical mundane considerations were the focus of discussions at weekly field meetings. At a meeting in mid-May, discussion centered on basic issues of coordination. How many people would be needed to do the work in the field? A number of undergraduate students had been hired, and certain people in the lab could be relied on to help on occasion. Weeding is a central part of the job of these workers. Before planting, sites are treated with herbicides, but thereafter herbicides cannot be used without contaminating experiments. Weeds compete with other crop plants for water and other resources and must be removed.

Questions of the effects of the weather on planning were prominent as well. Lab workers considered how long it would take to do the laboratory analysis of the soil from soybean roots after the harvest. In an early meeting, participants discussed whether or not to randomize harvested and non-harvested soybean plants, and how non-randomized picking might affect results of research on the ecology of the soil surrounding the plant roots. Plans were also made to take photographs of the fields.[6] Handelsman said that pictures dramatically illustrating treatment differences could be useful. In another instance, discussion centered on whether harvesting "dry weight" or "wet weight" was more appropriate. As the season wore on, these kind of considerations were replaced by discussions concerned with assuring appropriate staffing and more general updates on the progress at each of the field sites.

The management of resources is also a central part of life in the Handelsman lab. Students and staff are quite aware that the research they do costs money, and consequently, most are careful in using research materials. An observer might hear occasional discussions in the halls and whispers in offices when some lab members are irritated with others for neglecting the care of equipment or for being profligate in their use of supplies. Discussions in lab meetings are punctuated with announcements and questions concerning how much money is available or how much something costs. Funds are limited, and choices must be made. In chapter 4, I discuss a fatty acid analysis test that a Ph.D. student wanted to run on some of his *Bacillus cereus* strains. The student had to limit the number of strains he tested, because the cost of the analysis was prohibitive. A postdoc once proposed a test that might aid in a Ph.D. candidate's work on the so-called resistance gene. While he thought the test would provide convincing evidence, he noted that it was expensive. In these meetings, students sometimes ask Handelsman if funds are available for equipment, supplies or personnel. Similarly, Handelsman announces that resources are available for a certain item or that money must be spent within a limited period and seeks lab members' advice on how to use it. New computers were purchased after one such discussion, and supplies are stockpiled as grant terms come to an end.[7]

Related to the issue of funding is maintenance of equipment in the lab. Lab members are assigned specific pieces of technology to oversee.[8] If there are problems with a given tool, others in the lab know whom to contact. Often problems with equipment are aired in lab meetings. Because workers from both the Handelsman lab and the Goodman lab use the equipment, problems are not uncommon. One week, a lab technician announced that people were improperly adjusting the camera used to take photographs of fluorescent electrophoresis gels. Another week, a student carefully explained how to operate the vacuum pump because it was clear to him that it was being misused.

Of course, keeping the lab operating—providing the students with stipends and scientists with salaries, maintaining supplies, purchasing and sustaining equipment—takes money, and an observer's conversations with Handelsman would likely reveal that this is a constant worry for her. In a log she kept for me, Handelsman included virtually no commentary on the tasks that occupied her time, with one exception. A mid-week late evening entry said: "Worked on lab budgets, got depressed." In another context, I noticed how animated discussions with colleagues from other universities about the science underlying possible collaborations could quickly turn into discussions about how to fund the work.

Handelsman explicitly tries to insulate students from the vagaries of the economy of financial patronage, and my interviews with students in the lab suggest that her efforts have succeeded. Few students appeared to have a detailed understanding of how the lab is funded, although several noted that Handelsman sometimes announces when the lab receives a grant or a gift. Some students have fellowships, and they, of course, know where their support comes from. One postdoctoral fellow understood clearly that his support was precarious, and this was a source of anxiety for him. The one associate scientist in the lab was something of an exception in this insulated environment. As a post-postdoctoral scientist attempting to develop some professional independence, she was involved in writing and submitting grant proposals.

Overall, the operation of the Handelsman lab seemed unusually democratic for an academic lab filled with apprentice scientists. The advice of lab members was sought on matters both large and small. Still, Handelsman is clearly in charge, and although the students' futures obviously depend on her, it is also true that her future depends on them. In the end, if the research does not progress, if the experiments do not succeed, if the publications and patents do not continue, neither will the funding. In this context, the oversight of student work is crucially important. There is nothing heavy-handed in this oversight, however. Weekly meetings, visits in the lab, and chats in Handelsman's office, all give the lab leader a clear impression of what is going on.

In addition, once during each academic term, a lab meeting is devoted to a discussion of accomplishments and goals. Each lab member takes about five or ten minutes to discuss work completed over the last academic term, as well as what was on his or her agenda but was not completed. In addition, the scientists discuss in varying degrees of detail what they hope to do over the next term. It was commonly the case in the two such meetings I attended that the lists lab researchers provided detailing what they had hoped to do but had not accomplished was extensive. At one point in one meeting, Handelsman noted the "confessional" character of presentations. But one can imagine the goad to productivity such confessions provide. Several members of the lab told me that they felt guilty if they did not work "hard enough." If these confessions reinforce such guilty feelings, workers might very well redouble their efforts. In the end, at a professional level this is perhaps better for everyone concerned. If scientists at the bench work hard, their research is likely to progress, publications (and patents) are likely to follow, patron support is likely to be maintained, and students are likely to get degrees and jobs.

Handelsman's commitment to scholarship and education notwithstanding, she and her lab do have a relationship with members of the world of commerce. Soon after my arrival, members of the lab had lunch with two scientists from one of the nation's largest agrichemical firms. The exchange was informational. This was the only meeting of this kind that occurred during my time in the lab. According to Handelsman, members of her lab meet with representatives of this company twice a year.

More common, but still relatively infrequent, are meetings between members of the lab and representatives of firms that are collaborating with the lab, and that are providing support for some of the lab's research. During the summer, some of these meetings take place at the sites where the lab is undertaking field tests. I attended meetings in preparation for such site visits, and joined Handelsman and others on one tour. The first tour was to include marketing personnel, and Handelsman told lab members preparing for the visit, "This is not about science." Instead, the aim of the field tour was to provide company representatives with a visual sense of the effects of UW85 in contrast to other treatments. In contemplating the tour, careful attention was paid to the choice of field sites. Tables of data were to be placed on an easel for firm representatives to see, and Handelsman told the two lab workers coordinating the visit to make it "really simple. Just a few numbers that . . . marketing people can understand."

Later in the summer, I joined Handelsman, Goodman and two lab workers who were coordinating the summer field studies on a trip with two company representatives to view some of the fields in which UW85 was being tested. A lab technician who was partially responsible for coordinating the field trials told me some days before we left that preparing for this visit would probably take a day and as a consequence would require postponing alfalfa harvests at two sites. Interestingly, the company representatives were quite aware that it is problematic to make assessments of the efficacy of UW85 on the basis of "eye-balling." Nevertheless, when I asked one why they wanted to see the field sites when UW85 would ultimately be assessed quantitatively in terms of yield and emergence, he told me that he and his company have "gotta buy into the technology." Apparently, Handelsman and her university collaborators were right to focus their preparatory discussion on appearance, though it does not ultimately correlate with performance. There was, however, no effort to deceive the visitors. Indeed, Handelsman repeatedly pointed company representatives to cases where it appeared that the commonly used chemical was outperforming UW85.

In addition to occasional meetings with company representatives, members of the lab meet among themselves and sporadically with representatives of the Wisconsin Alumni Research

Foundation (WARF), which is the University of Wisconsin's patent agent, to consider intellectual property matters. I attended one of these meetings. Talk at this session did not involve the nitty gritty details of outlining a patent application. The aim appeared to be to give students and others in the lab some sense of what is patentable, and the kinds of inventions in which the Foundation is interested. The WARF representative spoke at a general level about these issues and the organization's patenting track record. Lab members asked questions and spoke about specific developments in the lab. This discussion revealed that these developments were not yet ready for patent application.

One could miss a few days in the lab over six months and easily miss the infrequent references to explicit connections between the Handelsman lab and the world of commerce. But I came to realize that the kinds of direct and ad hoc effects that influence university science—those effects that are often the focus of students of university-industry relations (e.g., restrictions on the free flow of information demanded by industrial funders of university science)—only tell part of the story. In addition to these direct effects, I discovered there are indirect, systemic effects of the commercial world on university science; I would suggest that they are just as important. If the Handelsman lab is representative, then these factors are transforming the everyday practices of university biologists, and no amount of good will or self-consciousness will allow university scientists to entirely escape them.

Let me put this a bit differently. I left the Handelsman lab with no doubt that it is a center for vibrant scholarship. But even though much of what happens in the lab is what laboratory members call "the science," Handelsman scientists are inescapably part of a world that is dominated by commerce. And indeed, precisely because the Handelsman lab is not the kind of place where scientific integrity seems compromised by big-dollar deals with high technology firms and consequent agenda tampering and secrecy agreements, if this lab—where the leader is deeply committed to traditional academic values and works hard to protect her students from the most troubling aspects of commercial interest in university biology—is affected by the world of commerce, then it may be fair to conclude that the influence of industrial culture and values on university biology is more subtle and pervasive than observers and participants typically realize.

Romancing a Foreign Culture

My relationship with Handelsman and other members of the lab was strained at times because of my view that indirect and systemic effects of the world of commerce on university science are indeed a central issue, as well as my contention that understanding this issue requires what could be understood as a structural analysis. My effort to understand the tensions between us provides an appropriate conclusion to this chapter and a useful foundation for the elaboration of my theoretical orientation, which I outline in the next chapter.

The role of ethnographer is a complicated one, and I want to touch briefly on that experience because it aids in explaining the conceptual orientation I take throughout this book. Assumptions and stereotypes are central to social life. Daily interactions would be impossible if we could not automatically and unthinkingly draw on knowledge of situations and roles to guide us in our actions and interactions (see Bourdieu 1984). I entered the Handelsman lab with expectations; likewise, lab members made assumptions about what a sociologist might think and do in their work environment. Furthermore, we all have self-images—senses of who we are and what we do. The self-images of lab members and my understanding of my objectives for this project at times conflicted. During my ongoing relationship with the scientists in the Handelsman lab, the assumptions held by all parties were challenged. This created some discomfort for me, as I expect it did for members of the lab.

Professor Handelsman made only one requirement of my participation: that I learn what it "feels like" to do science and to be a scientist. In exchange for agreeing to learn a few simple techniques and to diligently and repeatedly undertake a simple experiment, Handelsman gave me unrestricted access to her lab. I could talk to anyone, read anything, go into any drawer, attend virtually any meeting.

I cannot know for certain the assumptions that underlay Handelsman's conditions, but in retrospect it seems to me she must have thought that "getting the story right"—providing an accurate portrait of her lab—depended on my coming to understand the central component of what lab workers do. That understanding required that I do what lab workers do—that I feel the frustration of repeated experimental failure and the excitement of success. I did, at some level, learn what it felt like to do laboratory biology. I came to identify with the lab's successes and failures and to develop emotional bonds with members of the lab. Looking back, it certainly seems plausible that any disappointment or even betrayal that members of the lab felt after reading my work or hearing me talk about it could have been magnified because of the friendships that had formed among us.

My first few weeks in the lab were devoted almost exclusively to learning how to perform a particular test. I was taught how to determine whether or not a set of soil samples from Costa Rica contained a gene that codes for resistance to the antibiotic zwittermicin A. I worked with a senior lab technician, learning to separate the bacterial material from the soil. She taught me how to grow bacterial colonies on "plates." Following her instructions, I mixed up the growth medium. When it solidified in petri dishes, I dipped sterile toothpicks into water containing invisible bacteria. I spread the water on the dishes and then put the dishes in the "oven"—an incubator—overnight.

At the beginning, I left the lab at the end of each day with a headache. I felt the way I remembered feeling each evening while I was traveling in Spain many years ago. I knew the rudiments of the language, but each conversation was an effort. I was constantly engaged in translating in my head, and felt absolutely exhausted as I got into bed after each arduous day of making small talk, asking directions, ordering food, and planning activities. I came home from the lab feeling like that, and badly in need of a beer or glass of wine. My mind felt blurred and foggy—spinning with new information. I had no framework for organizing all that I was learning, but I was excited to return each day. I rushed my daughter off to school, parked my car in front of my house and quickly walked the three-quarters of a mile to the lab. Had I missed anything? I looked in the incubator. Where the day before there had been apparently nothing but a gelatinous growth medium, now on top of that there was a scribbled pattern of white fuzz.

My teacher in the lab showed me how to distinguish various types of bacterial colonies by appearance—their morphology. It was *Bacillus cereus* I wanted. My next step was to thin out the scribbles of fuzz by using sterile toothpicks to put less on a new plate, letting it grow in the incubator and repeating the exercise until there were clear and distinct little mounds of fuzz: bacterial colonies. Ultimately, I took a small plate divided into a grid on which each space corresponded to a numbered map of the plate. I touched a distinct colony with a sterile toothpick and then touched the toothpick to a spot on the grid plate. I repeated the process until I had four such plates. All went into the incubator overnight.

I returned the next day excited to see what I had: little dots—bacterial colonies—spread evenly across each plate. I stored these plates in a refrigerator filled with other plates, each with a lid, each taped shut, each piled neatly one on another. My plates had my initials on the sides. In the cool refrigerator, the colonies would grow slowly; it would be some time before one began to blur into another. If this got close to happening, I had to repeat the exercise to keep my "isolates" *clean*. Similarly, there was always a danger of some unwanted contaminant: a bright colored or hairy blob on a plate, threatening the integrity of my *pure* samples: foreign cultures. If they threatened my colonies, I would use the colonies from my grid plates to create new plates with fresh toothpicks and growth medium.

This part of my project is a central ritual in bacteriological research. As I was engaged in this task over my first several weeks in the lab, other lab workers were at their benches doing what looked like much the same thing. I had my own bench space, and as I sat patiently "patching" my colonies, lab members would engage me in conversation and offer words of encouragement and advice. The lab members were exceedingly welcoming. I could walk up to virtually anyone's bench and ask for assistance.

Though it is a central practice in bacteriology research, developing my own set of bacterial isolates was only the first part of my project. When my plates were prepared, my tutor taught me how to use the Robocycler to run PCRs.[9] By heating the double helix structure of a piece of DNA, the twisted strands are separated into two individual strands. So-called primers—short single stranded molecules—bind to the ends of the now single strands of DNA, and a DNA polymerase fills in the complementary nucleic acids in the remaining positions in the sequence; this creates a new double helix. This process is repeated, permitting the geometrical amplification of the target DNA fragment. The amplified product of PCR can be used in a wide range of further experiments.

I learned to mix several chemicals together in micro-tubes—very small plastic tubes into which I put extremely small quantities of each substance using a "pipetman" (a precisely calibrated tool for dispensing liquid)—under a sterile "hood." Each tube was placed in the Robocycler, where it was repeatedly heated and cooled over several hours. At the end, the result was supposed to be multiple copies of the gene that codes for resistance to the antibiotic zwittermicin A (Silo-Suh et al. 1994). I learned to test for the presence of this gene—"the resistance gene"—by running an electrophoresis gel. I first made the gel using an array of powders and water. I placed a "comb" in the liquid before it solidified. When it was solid, I pulled the comb out, and it left small holes in the gel into which I could pipet material from the tubes I had run in the Robocycler.

Electrophoresis permits one to detect the presence of genetic material by running an electric current through the gel. The current forces material of different weights to different places on the gel. In one lane there is "ladder," which serves as a yardstick, highlighting locations on the gel associated with different molecular weights. Since people in the lab knew the weight of the resistance gene, it is theoretically a simple matter to locate it in any of the lanes against the ladder. After the gel is run, I would dye it, and then photograph it under an ultraviolet light. Little lines matching various points on the ladder would indicate the presence of genetic material. Lines that appeared brighter in the photograph would indicate the presence of more genetic material. If any of my isolates had the resistance gene, they should have appeared brightly lit at the appropriate place. My first run indicated some lines or "bands," and I was excited, but there was contamination too. One lane into which I had put drops of a mixture containing water and other substances, but which contained no genetic material from my *Bacillus*, revealed several bands.

On my second try I got nothing. No bands at all. After several more gels with apparently no genetic material, I felt distressed, and lab members comforted me with stories about how temperamental and inconsistent PCR can be. After one of my unsuccessful attempts, there was some discussion of a joint meeting between Professor Handelsman's lab and Professor Goodman's adjoining lab to discuss what was going wrong with PCR. It was clear that my work was being taken seriously; I wasn't being treated as a novice or a guest. People tossed around several possible explanations for my problems. One researcher pointed to an article that discussed what can happen to mineral oil used in PCR if it is left under fluorescent light.

I tried a new bottle of oil. I tried all sorts of things. Then my tutor showed me a trick I had overlooked the first time she taught me how to do PCR. I needed to swirl the material in each micro-tube with a quick turn of my wrist.[10] When I did that, I started to get results. "Bands" of DNA appeared on my gels. One postdoc—who had had his own difficulties with PCR—started to call me "Mr. Bands." However, there was still contamination in my water lane, and although I repeated this test many times, I was never able to get consistent results.

When I started working on my resistance gene project, Handelsman mentioned the possibility of publishing my research results. It was another example of my work being taken seriously; it had the effect of making me feel invested in the lab's research effort. Now I wanted to succeed for me *and* for the lab. I imagined leaving the lab as a sociologist who also had a publication in plant pathology. It never happened, but nevertheless, I became increasingly drawn into the drama of laboratory research. I strained to understand results discussed at lab meetings. I stood in the hall with grad students and postdocs as they talked breathlessly about new findings or complained about hard work and no payoff.

I was quickly included in lab culture and was asked my views on music and university politics and encouraged to join lab members on sojourns to the local ice cream shop. Away from the lab, students engaged in friendly taunting of one another and freely complained about absent lab members. In addition to bench-top rituals, I was inducted into the lab ritual of Thursday afternoon Chinese take-out. Lunch was ordered in the late morning. A student first went to the bench of one of the lab's senior Ph.D. students where the left-hand drawer contained the take-out menu. The student in charge that week provided a sign-up sheet. After the last call, he faxed in the order. When the food was delivered, everyone, including Handelsman, would congregate in the crowded "community room," which soon filled with jokes and laughter. Handelsman was often the butt of good-natured jibes. Lab members imagined her as someone who must have been a precocious child. This, they speculated, explained her advanced standing in the profession despite her relative youth. Though young, her professional dedication (as well as the absence of a television at her house) meant that she was unaware of cultural trends, and her ignorance in this area was a source of humor to lab workers in their twenties. At the same time, she was perceived as a member of an older generation, and sometimes her answers to questions of how things were in her youth also evoked smiles and laughter.

It seemed to me that my initiation and integration into the community was rather thorough. I too became the butt of some humor. Unlike the scholar who ventures into a truly foreign culture—the European anthropologist examining an African village or the white ethnographer studying a Latino community in the United States—I dressed like other people in the lab. I talked like them. I was like them![11] I was not out of place with a yellow pad in hand taking notes at weekly meetings. Even at my lab bench, taking notes didn't make me stand out, and in the office I shared, I had a computer just as some other lab workers did. Still, while they accepted me, they knew I wasn't truly one of them. They made jokes as I took notes: "Quiet, Danny's listening," they laughed. They were curious. One person assumed I was studying "the social dynamics of the lab." Another told me as we left the lab together that he understood I couldn't tell him what I was doing, as that might "contaminate" my research.

In fact, I did tell people what I was studying, though in a general way. I gave Handelsman a research proposal, and early on she asked me to talk about my work at a weekly lab meeting. I described the work of Bruno Latour and Steve Woolgar on laboratory life and fact construction (1979).[12] I contrasted that micro-level study with my earlier macro-level work (1995).[13] I talked about the need to integrate these levels of analysis. I described my work in the lab as an attempt to gain some understanding of the ways in which the practices of the lab are affected by the world outside academe. I used papers published by lab members to describe the fact construction process as Latour and Woolgar might have done. This last point generated some apparent discomfort among people at the meeting. Handelsman told me afterward that I had misunderstood some of the work in the lab's papers I discussed. She warned me to be a bit more cautious but comforted me as well, indicating that lab members wouldn't hold my misrepresentation against me. And indeed, after feeling a bit awkward for a couple of days, all seemed forgiven.[14]

But the tension that emerged from my early talk recurred on other occasions when I presented my sociological research. My observation of the laboratory as a full-time participant ended in September of 1995. Before I left, a professor who organizes a departmental lecture series asked if I would be willing to discuss my project before the entire plant pathology department. I agreed, and we set a date for December. At this stage, I had only done the most preliminary processing of the data I had collected during my period in the lab. Still, I prepared a talk. In it, I discussed the need to fuse the kind of work done in laboratory level studies, like Latour and Woolgar's, with the kind of work done in the history of science and technology (see Kloppenburg 1988; Leslie 1993; Leslie 1994; Noble 1984), which examines the ways in which research patrons may shape research agendas, conceptual orientation, and ultimately experimental practice. I devoted part of my talk to discussing the intellectual property protection practices of the Handelsman laboratory and its relationship with the Wisconsin Alumni Research Foundation. In addition, I explored the problem of "fact construction" through multiple media. That is, while Latour and Woolgar examined the process through which scientific claims come to be accepted as facts through their circulation and alteration in scientific journals, I discussed a similar process that instead depends on multiple audiences—businesspeople, farmers, scientists, journalists, and newspaper readers—and diverse documents.[15]

In the Handelsman lab, there is a tradition for researchers to rehearse their talks before other lab members in advance of formal presentation. The trust Handelsman lab workers placed in me by opening their professional lives to my gaze made it seem appropriate that I should honor this tradition and do a dry run.

Prepared with the visual aids, I made my pitch before members of the lab in the Community Room. Smiles and laughs faded quickly as I spoke. Everyone was silent, and as I caught Handelsman's expression, I thought I saw distress. I was not at all sure what I had done wrong, but my audience was clearly unhappy.

At first, most everyone was quiet. Then in a measured tone, without a hint of anger, Handelsman explained that the portrayal that I presented would be misunderstood by her colleagues. They would view my representation as showing a lab that did substandard work, since one thing I drew attention to was the limited number of citations to the lab's work I found in a database search. My own view was that this fact said more about the organization of the research area in which the lab works than about the quality of the research they do (see Whitley 1984), but one could not be certain that my plant pathology audience would share this reading. Similarly, Handelsman expressed some concern that attention to intellectual property issues might make the lab seems greedy. In light of Handelsman's comments, I made some adjustments to the talk. The changes, which Handelsman and I discussed through e-mail, seemed to satisfy her, and I did not in any way feel that I had been forced to compromise my argument to placate my subjects.

Six months later, I began work on two papers based on my study of the Handelsman lab. When I had complete drafts, I sent them to Handelsman and suggested that they might be circulated to other members of the lab and to Bob Goodman as well. After Handelsman read one of the papers, we arranged a meeting. Handelsman and I always got along well—talking about subjects ranging broadly from science to politics to children. But when I entered her office, I felt uncomfortable. I thought she was short with me, if not angry, then disappointed or upset. After we began to talk, Goodman joined us.

Our discussion focused on my failure to properly circulate my drafts to members of the lab and to Goodman. Handelsman and Goodman concurred that I was wrong to depend on Handelsman to circulate the single copy I had sent her. Handelsman for the first time raised the issue of anonymity. Was it not possible and even appropriate to leave the name of the lab and lab personnel out of the paper? One portion of the paper discussed

Goodman and a project being done under his supervision. Goodman suggested that first, I did not have permission to write about his lab, and second, that I had misrepresented the facts.

I had clearly made several errors of judgment. I should have discussed the issue of anonymity prior to initiating the study. I should have formally asked Goodman if I could study his lab (the collaboration in the mid-1990s between Handelsman's and Goodman's labs sometimes made the boundary between them less than distinct), and I certainly should have circulated drafts of my paper more broadly. I believe we ultimately cleared these matters up,[16] but I suspect that our discussion in Handelsman's office in the fall of 1996 was about more than the issues we explicitly discussed. I expect that Handelsman and Goodman's discomfort, irritation, or disappointment arose at least in part from how they perceived *my* representation of *their* lives. My reading is reinforced by Handelsman's concern that my talk prepared for the plant pathology department could be misunderstood by the audience.

Further evidence in support of this interpretation can be found in the margins of a copy of one of the papers I wrote on the lab. The copy was returned to me with written comments by Handelsman and one of her graduate students. In a portion of the paper that discusses patenting issues, the student wrote what I took to be a sarcastic comment: "except for the *Bacillus* isn't everyone just after profit?" And indeed, Handelsman explicitly pointed out that my paper could give the impression that the everyday life in the lab is dominated by intellectual property concerns, when the truth is that such matters take up only a relatively small amount of lab members' time. I failed to convey how much time was spent in the lab doing experiments, writing papers, and discussing research (1998a).

This brings me back to issues of knowing what it "feels like" to do science, the lab worker's self-image and assumptions. The first papers I wrote and shared with members of the lab did not clearly reveal what people in the lab perceive as their primary jobs—what tasks occupy most of their time. They reasonably may have assumed that I would be attentive to this in my own work. In theory, the many hours I spent doing experimental work would have given me ample first-hand material for rich discussion of this. But there are other books that already cover this ground. Natalie Angier's *Natural Obsessions* (1988) is an excellent example. Angier observed work in two laboratories that were investigating the genetic workings of cancer. Her vivid, evocative study captures the excitement and heartache of cutting-edge research. She explores how one experiment leads to the next, how error leads to insight, how conflict between labs promotes solidarity, as well as how family life can clash with lab life. She humanizes the scientists in her study; these are real people with feelings, families, and feuds. Angier describes how one scientist's obsession with a coworker crippled his research (1988, 124). She tells us how another treated the lab leader's suggestions with transparent disdain (1988, 219). Another scientist in Angier's story wept upon learning that he had been removed from a project (1988, 261).

At the same time, Angier's often compelling study *romanticizes* science and the life of the scientist. She talks, for example, about her fascination at seeing purified DNA for the first time:

> "I can't believe it," I said. "It's so *clean*." I thought if I grabbed the DNA, I could roll it in my hands, twist it, pinch it, shape it into little animals, even drop it on the floor, and it would still jiggle merrily, holding life so firmly that it was, on its own, alive. . . . I think that watching DNA come out of solution should be required of every high school student. Then perhaps people would realize that DNA is not a set of balls and sticks or a diminutive staircase. Real DNA is like the sound you hear when you hold a conch shell to your ear. DNA is music. (1988, 35)

These people are doing something that is at once out of reach to most of us and truly amazing. These are extraordinary people. Of cancer researcher Robert Weinberg Angier says:

> I'm not a soothsayer, and it's foolish of me to make predictions, but I believe that when something really big breaks in cancer research, something that will benefit a substantial fraction of cancer patients, Bob Weinberg will, directly or indirectly, be involved with that breakthrough. He's intuitive, creative, always clever, and occasionally brilliant; the best young researchers in the world clamor to work with him. Weinberg is also one of the luckiest scientists alive. (1988, 43)

A scientist who runs a competing lab is similarly portrayed as extraordinary by Angier:

> Michael Wigler is a child prodigy who just happens to be an adult. Great scientists are always "bright," but Wigler has the balletic, flamboyant, tireless intelligence of a *Wunderkind.* He's more sheerly intelligent than many of his peers, and his peers will admit as much. (1988, 269)

Angier is critical of science writing that focuses on "done science," rather than "doing science," because such an approach underplays the hard work that science is. This approach to writing about science often emphasizes, in Angier's view, "the gee-wizardry of science, the 'spectacular' or 'revolutionary' discoveries that seem to spring to life parthenogenetically, with little or no evident effort behind them" (1988, 6, 7). Nevertheless, Angier's work resembles other science writing in a fundamental way. Outcomes and effects are explained in *individual* terms.[17] She pays little attention to the shaping and constraining influences of the larger environment in which scientists do their work. Angier assigns personalities a pivotal role in her explanations of the character of the laboratories (1988, 47). She points to idiosyncrasies in an individual's biography to explain how and why someone enters science (1988, 299). Competition within the lab arises because of the personality traits of individuals (1988, 260), and factors beyond the individual become an issue only in the competition between labs that drives research (1988, 102). Despite the fact that one of the central people in her study helped establish a biotech company, the commercial world hardly seems to exist in Angier's account (1988, 170).

I suspect that many scientists would be more comfortable having their work represented the way Angier portrays her two cancer research labs or even the way I portray the Handelsman lab in the first part of this chapter, than the way I depict life in the Handelsman lab in subsequent chapters. Individualist beliefs run deep in U.S. culture (Bellah 1996 [1985]). We learn to see the world as the product of individual conscious action. This orientation permits us to take credit for our success and to blame others for their failures.[18]

Each year in my teaching, I find it difficult to challenge this perception among my introductory sociology students. C. Wright Mills presents the problem nicely:

> Seldom aware of the intricate connection between the patterns of their own lives and the course of world history, ordinary men [*sic*] do not usually know what this connection means for the kinds of men they are becoming and for the kinds of history-making in which they might take part. (1959, 3, 4)

In this context, Jay MacLeod (1987) notes that Americans never tire of the mythic stories of Andrew Carnegie and Horatio Alger. Anyone can move from rags to riches. MacLeod calls this the "achievement ideology." If we do not successfully move up the social hierarchy, our failure is evidence of our own inadequacies. More generally, individualism is an *agency-centered perspective* that suggests we make ourselves and are not shaped by the complicated social matrices in which we are embedded. One can imagine that a person with an individualist or agency-centered outlook might find it disconcerting to see a portrayal of her world that attributes great importance to the impact of *structural factors* on her daily life. The portrait might seem to implicate the individual's motives, and if the representation is critical, that criticism might be perceived not as an assessment of some structure in which that individual exists, but rather as a condemnation of the person.[19]

Although I cannot know for certain, I suspect that at some level the problems in my relationship with members of the Handelsman lab arose out of this tension between individual and structural representations.[20] I was interested in understanding how structures that were beyond the control of any individual could shape the practices of lab workers. But as the student who made the comment that I seemed to portray scientists as greedy capitalists suggests, she perceived my claims about structure as personal—as statements about lab members' control over their own situation. In her mind, my approach may have suggested that I believed that the lab had consciously chosen to follow this "greedy" path. Similarly, Handelsman may have been concerned that the attention I paid to intellectual property issues in my paper would suggest to her departmental colleagues that there was an unrestrained commercial orientation in her lab, rather than a dedication to the higher ideals of science. And while I believe Handelsman and members of her lab are committed to the "higher ideals of 'pure' science," this is not the concern of my study. To reiterate: I am interested in how the practices of scientists are shaped by the larger world in which they operate, and the extent to which this world is beyond their ability to control it. Thus, this book is more about the world in which the Handelsman lab exists than about the lab itself.

Interestingly, I think Handelsman and at least some members and former members of her lab are not simple individualists or voluntarists who believe that all individual actions are based on personal choice. As I discuss in chapter 6, Handelsman shows herself to be an adept social analyst, and to have in C. Wright Mills's term a "sociological imagination": the capacity to understand how characteristics of the social world shape individual action. But in my experience, even sociologists who clearly perceive the constraining capacity of structures in their scholarship often discuss their personal behaviors as though they were unaffected by outside influences. It is difficult to take complete credit for one's successes if one applies a sociological imagination to one's own circumstances. It is easier to have a sociological imagination at a distance than to direct it at one's own practices.

I did have a *romance* with the science done in the Handelsman lab and with the researchers in it. They and their work are, indeed, extraordinary. But my aim was not to echo Angier. I believe this is one source of the tension between workers in the Handelsman lab and me. My goal from the outset was to collect data to enable me to explore how the world in which the lab is embedded may shape the practices of the lab, and particularly to see how and to what extent what I call the "world of commerce" influences those practices.

In the end, I hope Handelsman and her colleagues understand the aim of my work. It is not intended to impugn the integrity of these scientists. They are all people whom I have come to respect, with whose project I am fascinated, and whose goals I admire. This finally brings me to the title of this book: *Impure Cultures.* In selecting this title, I do not intend to suggest anything untoward about the work of the Handelsman laboratory or about the environment in which it is embedded. I use the term "impure cultures" as a metaphor, which is apt, I think, because it refers to both the difficulties of maintaining pure bacterial cultures in microbiological research (something I now know a great deal about) and the difficulties in maintaining rigid boundaries between the worlds of university biology and commerce. Of course, researchers actually can maintain pure bacterial cultures—indeed, they must. But social purity—maintaining the separation of distinct social milieu—is neither possible nor generally appropriate. My interest lies in investigating the nature and implications of this social blending.

Book Outline

In this introductory chapter, I attempted to do two things. First, I tried to provide a sense of the extraordinarily heterogeneous character of life in the laboratory; it is both the focus and starting point for my investigation. This portrait suggests that the world of commerce does not explicitly dominate the day-to-day lives of members of the Handelsman lab. I found no evidence of the kinds of violations of the norms of academic science that concern many analysts of university-industry relations. Instead, one sees in this lab a center of scholarship and education. This makes the Handelsman lab a complicated case to use in an effort to understand the commercialization of the university. We must, as I argue in the subsequent chapters, examine not only the explicit and overt effects, but also the more subtle, pervasive influences the world of commerce exerts upon university science.

In the second portion of the chapter, I reflected on my research experience and how difficult it is to travel a terrain where the distinction between subject and object blur. More specifically, I suggested that it was my efforts to understand the life of the Handelsman lab in structural terms that led to tensions in my relationships with lab workers. Finally, where the first portion of this chapter lays the foundation for a discussion of the connections I point to between the Handelsman lab and the world beyond it, the second portion provides entree into the scholarly debates (such as the issue of agency versus constraint) in which I engage and the position I take.

In chapter 2, I provide a detailed analysis of the literatures on which I draw and to which I speak. My work addresses two areas of study in detail: science policy analysis of university-industry relations, and recent work in science studies. Through my assessment of important work in these areas, I attempt to build a case for the approach I take in the subsequent chapters.

In chapter 3, I explore the history of biocontrol. I investigate the intimately linked history of this research area and the development of agrichemicals; I argue that industry may be able to shape university research even without *directly* funding it. Through the use of two case studies—insect control in the citrus industry and the development of UW85—I show that research practices can be *indirectly* shaped by industry.

In chapter 4, I examine an area that is not frequently considered in discussions of changes in the biological sciences: the commercialization of tools and supplies employed in university biology. Here, I explore three cases arising out of work in the Handelsman lab. I investigate a contentious relationship between the lab and a commercial supplier of research services, analyzing how a relationship of resource dependence overlaid with conflicting understandings of reputation shaped the interaction. This is followed by consideration of how the black-box nature—the largely invisible and unexamined character—of a commercially produced research tool affected where scientists looked for contamination. Finally, I explore the implications of the use of "kits" in life science research for the education of scientists-in-training.

In chapter 5, I take up a host of problems raised by attention to matters of intellectual property in one lab. Concretely, I consider three ways in which the practices of academic biologists—members of the Handelsman lab in particular—are shaped by the intellectual property regime in the United States. First, I investigate how corporate patenting of research tools can indirectly affect the everyday practices of scientists. Next, I analyze the unequal relationship between the Handelsman lab and the Wisconsin Alumni Research Foundation. Finally, I consider the influence commonly held assumptions and attitudes about intellectual property protection have in the university lab.

In chapter 6, I explore two very different efforts undertaken by Handelsman and her colleagues. The first involves a multidisciplinary collaboration that challenges a set of existing social relations; the second is the establishment of a research institute that capitalizes on increased industry interest in a particular research area. Pairing these two cases leads me to conclude that contrary to a widely held position in science studies, although the social and technical are integrally related, in order to understand how these two cases developed, it is profitable, for analytical purposes, to distinguish between the social and the technical, and, indeed, to examine the analytical priority of the social.

2

Traversing the Conceptual Terrain

The process of collecting and sifting through ethnographic data is inevitably shaped by an incalculable set of assumptions with which one enters the research site. This is an age old problem. It has been both ignored (for example, in the case of empiricism) and considered obsessively (for example, certain varieties of reflexive social science). There is no way for me to account for all of the factors that influence my reading of my experience in the Handelsman laboratory or my organization of the data I collected. I am aware, however, that my experience in the lab and my analysis of the data has been shaped by an ongoing dialog with three literatures: the policy scholarship on university-industry relations (UIRs), laboratory studies, and two varieties of agency-centered analysis in science and technology studies.

In the mid-1980s, I participated in a study of university-industry relations at the University of Wisconsin's College of Agricultural and Life Sciences (Kleinman and Kloppenburg 1988). In that project, we considered the concerns that were typical at the time in discussions of UIRs: corporate influences on the creation of university research agendas, corporate restrictions on the free flow of information, and questions of control over intellectual property. The data collected for that project consisted of university reporting forms, research contracts, and patent agreement documents. While undertaking that project, I was influenced by recent laboratory studies (see Latour and Woolgar 1979; Knorr Cetina 1981), and wondered how our analysis might have been different had we been able to spend time as participant observers in university laboratories. When I entered the Handelsman laboratory in 1995, I assumed that the problems I would address in writing about my laboratory experience would mirror those we explored in our case study in 1988. I found a vastly more complicated situation than I expected, and this led me to rethink the literature on university-industry relations. Thus, the first section of this chapter considers the UIR literature in light of the reevaluation of it I was forced to make as I spent time in the Handelsman laboratory.

As I stated above, it was both my involvement in research on university-industry relations and my reading of two seminal laboratory studies that led me into the Handelsman laboratory. Three of the four book-length laboratory ethnographies that are part of the first wave of such studies (Latour and Woolgar 1979; Knorr Cetina 1981; Lynch 1985) are largely or exclusively "micro" in their focus. That is, they treat the laboratories they study in relative isolation from larger social factors. In entering the Handelsman lab, I imagined that this kind of orientation would limit my capacity to explore the issues that I thought were important in the UIR debate, and once I was in the lab, it became clearer still that the concerns I came to regard as central could not be adequately addressed at a micro level. In the second section of this chapter then, I outline the major findings of several prominent laboratory studies and consider the ways in which my work both builds upon and diverges from them.

In the last section of this chapter, I explore work in the actor-network tradition (see Latour 1987) and the social-worlds approach (see Fujimura 1987, 1988). My confrontation with this work dates from my reading it while I also read work in institutional political economy (see Hall 1986; Lindberg 1982) and political sociology (see Evans, Rueschemeyer, and Skocpol 1985); at the time, I was unable to imagine that this agency-centered literature had the capacity to explain the phenomena that then concerned me (see Kleinman 1995). As I worked in the laboratory, I regularly "tested" agency-centered readings of what I saw with a more structural or organizational point of view, based on my understanding of institutional political economy and political sociology. I found the latter discussions consistently more compelling. In the final section of this chapter, I outline the framework upon which I draw to analyze university biology and the world of commerce in the chapters that follow.

The Myth of the Ivory Tower

The advent of genetic engineering has brought with it an extensive policy literature on UIRs.[1] A central assertion of much of this work is that the development of university-industry relations in the biological sciences in the United States over the past quarter of a century has served to fundamentally undermine the autonomy academic scientists enjoyed in their research practice prior to the development of these relationships.[2] Much of the literature in this area implicitly contrasts the contemporary scene with a "once-remote ivory tower"—the idea of the university as separate from society and untroubled by practical concerns.

In this section of chapter 2, I review some of the literature on university-industry relations. My criticism of this work is twofold. First, by assuming that the academy was once an isolated ivory tower, these interventions into policy discussion serve more to reinforce a powerful myth than to shed light on a complex process. Significantly, while the nature and extent of dependency has varied over time, geography, and type of institution, periods of relatively high levels of faculty autonomy regarding their capacity to define research agendas and set priorities are relatively few in the history of the American university. Second, I suggest that much of the research on UIRs has trained its attention on the impact of *direct relations* between academic scientists and science-based firms. In doing so, this writing largely overlooks the *indirect but pervasive* influences of the world of commerce on the daily practices of university biologists. In the chapters to come, I show that the Handelsman lab and others like it do not exist in an ivory tower, but rather in a dynamic academic environment in which the indirect effects of the world of commerce and the larger world in which academic science is embedded are substantial and largely inescapable. This conclusion has important implications for the direction public policy should take.

One can see the premise that the new UIRs constitute a novel breech of the "once-isolated ivory-tower" woven throughout the policy literature on university-industry relations in the biological sciences. A 1983 report by the American Association of University Professors (AAUP), for example, noted a "much expressed" fear "that university scientists may be pressured into taking work on research problems that do not interest them by a university eager to acquire a profitable patent, or to please or attract a corporate associate" (1983, 21a). According to that report, "A related fear is that a university may allow its corporate associate to interfere in a faculty member's choice of research topic" (1983, 21a).

Similar worries were expressed or documented in other periodicals. A 1989 article in the *Bulletin of the Atomic Scientists* noted critics' complaints that the marketing of university research products had "created a secretive and repressive atmosphere that threatens traditional academic freedom" (Hart 1989, 28). Taking a slightly different angle on the problem, but relying on the same underlying ideas of scientist autonomy, Krimsky warned that UIRs threaten to undermine the existence of a "disinterested intelligentsia to whom the public can turn for critical evaluation of technological risks, goals, and directions of biotechnology" (1984, 4).

The now classic study by Michael Blumenthal and his colleagues (1986a and 1986b) translated these kinds of concerns into hypotheses, and sought through a national survey of academic life scientists and biotechnology firms to provide empirical ground for debate. Blumenthal and his associates found that 30 percent of faculty members who received industrial support said that commercial considerations influenced their project choices to some extent or to a great extent, while only 7 percent of those without such support said commercial factors influenced project choices (Blumenthal 1986a, 1364). Faculty with industry support were also four times more likely than those without corporate funding to acknowledge that their research had resulted in trade secrets (Blumenthal 1986a, 1364).

In a 1990 follow-up to the Blumenthal study, James Curry and Martin Kenney used Blumenthal's survey instrument to explore concerns of agenda setting, secrecy, and the like at land grant universities. They found that over 40 percent of researchers receiving industry support acknowledged that commercial considerations affected research project choices. Significantly as well, over a quarter of those without industry support reported that commercial factors influenced project decisions (1990, 52).

The concerns that motivated early policy debate and scholarly research have not disappeared. In their 1998 study that surveys the existing literature and draws on their own work, Lewis and Seashore explore the threat to traditional norms posed by the spread of UIRs (see also Lee 1996). They conclude that "significant changes in the faculty-institution relationship are occurring . . . [and] as research relationships come to have the character of governmental or corporate contracts, the special norms of autonomy and self-regulation which have distinguished academic work in the past tend to have less certain status" (1998, 90). Finally, a 1999 article in *The Nation* reiterates concerns that university-industry partnerships are "eating away" at the "soul of academic science" (Shenk 1999, 12). The article's author, David Shenk, recounts stories of scholars who received industry support being prohibited from publishing research findings, and cites bioethicists who worry that university scientists who profit from their own research may "be tempted (consciously or unconsciously) to design studies that are more likely than not to have an outcome favorable to the product" (1999, 12).

The literature on UIRs does reveal important phenomena: the role of industry in setting agendas in academic science, the effect of academic relations with industry on open communication, and the ways in which UIRs can diminish disinterested expertise. But these are not novel threats. In one way or another, university patrons have affected research agendas and priorities since the beginnings of the research university in the United States. Historians commonly date the origins of the U.S. research university to the Civil War years (Gruber 1975; Bruce 1987), with research-based graduate studies coming soon afterward (Geiger 1986, 9). In these early years, a mix of sources including tuition, ad hoc philanthropy, and government funding supported university research (Geiger 1986). But more than a quarter century before the Civil War, firms occasionally employed university scientists to do research (Noble 1977, 110). Just after the turn of the twentieth century, the University of Illinois established the first engineering experiment station in the

United States—an organization that a dean of that institution regarded as an embodiment of a recognition of the university's responsibility to industry (Noble 1977, 131, 132). Similarly, by the 1920s, special institutes had been established at some universities to provide "industry with applied research and development services" (Geiger 1993, 297).[3] For example, the state legislature authorized the establishment of the semi-autonomous Engineering Experiment Station at Georgia Tech in 1919 (Geiger 1993, 284; McMath et al. 1985, 189). And indeed, the establishment of Georgia Tech in 1888 was meant to create an institution that would have "stimulating effects on the industrial economy of the south" (McMath et al. 1985, 35).

Industrial patronage did not dominate university life in the early twentieth century, and in the 1920s and 1930s, when development of the U.S. research university was increasing, most support came from private sources: foundations and private individuals (Geiger 1986; Geiger 1993, 45, 92). But private sources of support never meant absolute autonomy. Robert Kohler (1990), for example, traces the role played by the Rockefeller Foundation in the promotion of applications and techniques of the physical sciences to biology beginning in the early 1930s. Foundation funding went to scientists willing to utilize the physical sciences in their biological work. In this way, Rockefeller patronage shaped research agendas and the direction of entire fields of research.

As many commentators note, both World War II and the Cold War played crucial roles in shaping research universities in the United States (see Geiger 1992; Geiger 1993; Leslie 1993; Leslie 1994; Lowen 1997). World War II led to a massive transformation of the U.S. academic research economy in the United States. Beyond the research undertaken directly by the military during the war, the federal government contracted with universities to undertake research and development on devices and mechanisms of warfare through the National Defense Research Committee and later the Office of Scientific Research and Development (Kleinman 1995, 52–73). These contracts laid the foundation for federal patronage relationships with universities after the war's end.

From a relatively limited role during the war, the federal government came to dominate patronage for academic research in the years after the war (Geiger 1993, 48). According to Roger Geiger, the ideal of "unrestricted support" for academic research became "increasingly elusive" after WWII, as federal patrons stressed "focus and accountability" (1993, 49). After 1950, there was growth in the support for primarily "applied or highly programmatic projects." Three-quarters of federal support for academic research came from the Pentagon and the Atomic Energy Commission (Geiger 1993, 157; see also Geiger 1992, 26, 27).

Stuart Leslie succinctly captures the pervasive impact of military patronage on engineering and physics during the Cold War: "Predictably, given where the money was coming from, those topics directly relevant to the missile age—ultrahigh-temperature structures and materials, inertial guidance, hypersonics—grew robust, while those with strictly commercial applications—improved safety, increased fuel economy—languished" (1994, 216). The effect of military patronage on electrical engineering in the postwar period is clear as well. Leslie contends that the character of the knowledge produced was itself shaped by military patronage. According to Leslie, "Whereas questions about alternating current, radio, and high-voltage power transmission dominated the prewar curriculum, problems in microwave and solid-state electronics, communications theory, and plasma dominated the postwar curriculum." (216, 217).

The Massachusetts Institute of Technology (MIT) and Stanford University provide two clear cases of the impact of Cold War military support on academic institutions. At MIT, research policy was determined primarily by research patrons, and in most cases this was the Department of Defense (DoD) (Geiger 1993, 65). In the period after the war, the joint services provided substantial support for MIT's Research Laboratory of Electronics (RLE). In return for this support, which the Institute could "allocate as it saw fit . . . , the services wished 'to maintain close liaison between the military and the frontiers of electronic science and engineering'; and to have 'a laboratory from which the military services [could] . . . draw competent technical help at critical times.'" (Geiger 1993, 66). In 1946, the U.S. Navy asked the Laboratory to develop a missile guidance system, and in 1950, at the start of the Korean War, the RLE was asked to work on a number of explicitly military projects, including an early warning strategic radar system (Geiger 1993, 66, 67).

At Stanford, as elsewhere, a range of factors laid the groundwork for the postwar character of the university, including alliances established during the First World War, the economic conditions during the Depression, and feelings of patriotic duty during the Second World War (Lowen 1997, 32, 44). The First World War "brought together government officials, representatives of industrial enterprises, and leaders of America's scientific community" (Lowen 1997, 21). During the war, the National Advisory Committee for Aeronautics (NACA) set the aerospace research agenda at Stanford over the objections of the university's aeronautical engineers (Lowen 1997, 46). But the military was not the only patron that shaped research agendas and subsequent practice at Stanford. During the period between 1944 and 1946 Stanford administrators made efforts "to create institutes and other organizations to attract industrial patronage, and to reorient particular university departments to serve better the interest of regional industry, particularly aeronautics, electronics, and oil companies" (Lowen 1997, 75). Part of the motivation for this drive was to create some balance, preventing military-sponsored research from dominating the university (1997, 97). The Stanford Research Institute was created in 1946 specifically to undertake industrial research (1997, 77); the microwave laboratory was similarly established at the war's end as a means of maintaining relations with industry (Geiger 1993, 120, 121).

Just as it had at MIT, the Korean War brought an increase in military support for research of an "applied or highly programmatic nature" to Stanford (Lowen 1997, 121). According to Lowen, the applied research programs at Stanford "were subject to more direction from military sponsors than was the basic research and to the imposition of production schedules" (Lowen 1997, 121). Such relationships reinforced academic-industrial relations as the military imposed pressure for "quick development and production of new war technologies" (1997, 121).

During the Cold War years at Stanford, Lockheed selected and provided partial financial support for two aeronautical engineers, which led to a shift in orientation of the university's aeronautical engineering program from commercial to military (1997, 134). Beyond this, administrators sought to attract new faculty members whose research interests were consistent with the military's interests (1997, 138). In short, according to Lowen, "during the 1950s and 1960s, academic departments, and Stanford as a whole, became financially dependent on those whom they served—primarily military patrons, but also other federal patrons as well as private foundations and some industrial concerns" (1997, 188).[4]

According to Roger Geiger, there was a brief period that in retrospect may have appeared to be a "golden age" for academic science: the period of the mythic ivory tower (1993, 174). The launch of Sputnik in October of 1957 spurred federal support for academic research. Switching from a commitment to supporting research of an "applied" or "programmatic" nature, federal policymakers articulated a need to support "disinterested" research, and by 1968 federal funds were heavily weighted toward support for this sort of scientific investigation. It is worth noting, however, that this shift did not mean a decline in support for research that met specific programmatic needs of the government. Instead, the overall research budget increased substantially, allowing the "basic" portion to proportionately dominate money for "applied" research (Geiger 1993, 166). From a $456 million budget in 1958, federal spending for academic research jumped to $1,275 million in 1964 (Geiger 1993, 173).

Commitment to this distribution faded by the late 1960s, and even the National Science Foundation, founded specifically to support so-called basic research, was called upon to fund "applied" research as well (see Kleinman 1995; Kleinman and Solovey 1995). After 1968, federal funds were directed to "applied" university research at the expense of "basic" investigation (Geiger 1993, 196), and according to Geiger, "Instead of setting their own agendas, the research universities of the 1970s were placed in the position of defensively shielding their fundamental purposes against unsympathetic or hostile critics" (1993, 252). Outside of the universities, patrons and critics wanted something tangible for their money. During this period, "some departments admitted moving into applied research when funding for basic research was not forthcoming" (Geiger 1993, 272),[5] and one report commented on the apparent increase in university-industry relationships (Geiger 1993, 273).

This history makes it clear that it is simply incorrect to assert that the new university-industry relations of the late-twentieth and early-twenty-first centuries constitute a novel threat to autonomous faculty control of research agendas and priority setting. It is problematic as well to assert that these partnerships mark an exceptional incursion into the idyllic free exchange of ideas and research materials. Recent research suggests that even when commercial considerations are not at stake, information and research materials may not flow freely (Kleinman 1998b, 98). Inter-lab competition often leaves researchers reluctant to supply research materials, and inadequate maintenance practices may make it impossible to provide requested biologicals.

A survey of life scientists in the late 1990s found that 24 percent of respondents said that financial interests or agreements with a company affected their decision to withhold information. But almost twice that many respondents said that they withheld data and/or materials from colleagues to protect their scientific lead; another 27 percent said that cost affected their decision to restrict the flow of data and/or materials from their lab (Marshall 1997, 525).

A survey of academic geneticists undertaken in 2000 revealed similar results. In that study, 27 percent of respondents said they withheld post-publication information, data, or materials to honor the requirements of an industrial sponsor, and 21 percent explained their unwillingness to provide materials and/or data as stemming from the need to protect the commercial value of research results. But those percentages are low compared with some of the other explanations these geneticists gave for withholding information or material. Eighty percent cited the effort required to produce materials or information, 64 percent pointed to the need to protect a student's or colleague's ability to publish, 53 percent said they needed to protect their own ability to publish, and 45 percent cited cost as the reason they withheld data or materials (Campbell et al. 2002).

Finally, the idea that in that past citizens have been able to turn to the university for wholly objective and disinterested advice must also be challenged. While the nature of disinterestedness surely varies by time, place, and field, as recent work in science studies clearly shows, science is never wholly disinterested. That is, "non-technical" considerations commonly affect the practice of academic science. The ways in which investigators approach research problems are always affected by factors that in turn shape the conclusions at which they arrive. Whether it is the influence of research group affiliation upon one's view concerning what counts as a competent experiment (Collins and Pinch 1993), how pressure from colleagues creates reluctance to voice unpopular views publicly (Goodell 1979; B. Martin 1999), or the ways in which professional norms shape criteria for accepting results (Brown and Mikkelsen 1990), scientists can never provide anything we could call disinterested guidance.

In presenting this portrait, I do not mean to suggest that the economy of university patronage has not changed in recent decades. During the Reagan administration, the National Science Foundation began to provide funding for university-industry collaboration, and during this period, the fastest growing component of university research support came from industry (Geiger 1993, 305, 311, 315). A substantial number of the university-industry research centers funded jointly by federal and state governments, industry, and universities were founded in the 1980s. Indeed, of those in existence in 1990, just under 60 percent were founded in the 1980s (Brooks 1993, 218).[6] Commentators maintain that, in addition to support from industry in an environment of international economic competitiveness, the United States and other national governments shifted resources to economic development from social welfare. This means that the trend toward support of "targeted or commercial or strategic research" over so-called basic or curiosity-driven research continued unabated (Slaughter and Leslie 1997, 15, 37).

What is more, financial support for academic science with commercial potential was reinforced in the 1980s by a set of policies aimed at facilitating efforts to get science-based products to market. The Bayh-Dole Act of 1980 "signaled the inclusion of universities in profit making" (Slaughter and Leslie 1997, 45; see also Slaughter and Rhodes 1996). Prior to the act's passage, universities needed to secure government approval—no easy feat—to obtain patents on research funded by the federal government. Other legislation, like the Stevenson-Wydler Act, also passed in 1980, facilitated technology transfer between private and public entities. And measures like the Omnibus Trade and Competitiveness Act of 1988, along with a series of court decisions, bolstered the status of intellectual property, a crucial component in the commercialization of academic science.

Although industry support is still a relatively small portion of total university budgets, overall we have seen an increase in industry's contribution to academic science in recent years. While Hackett (1990, 251) suggests that under 3 percent of total academic R&D was supported by industry in 1972, a 1998 study contends that some 8 percent of all university research in the United States is funded by industry (Etzkowitz and Webster 1998, 27). The latter estimate may be too conservative, for an earlier study put the figure at about 12.5 percent (Blumenthal et al. 1986b). Of course, there is variation by field. In biotechnology in the 1980s, for example, industry provided over one-third of all support for university-based research (Blumenthal et al. 1986b).

If a transformation of the university is underway and the influence of industry on academic life is growing, it is important to understand that *the direct impact today of industry on academic science through patronage and explicit partnerships is part of a long historical trend.* The American university was *never* an ivory tower. Furthermore, to adequately investigate the distinctiveness of the current changes in the U.S. university, we must be attentive not only to direct factors, but also to the indirect and pervasive effects of such factors as the historical legacies of research fields (a central topic in chapter 3), the standardization and commercialization of research tools (chapter 4), and the character of the U.S. intellectual property regime (chapter 5) (see also Kleinman and Vallas 2001).

Laboratory Studies

The set of so-called laboratory studies published between the late 1970s and the late 1980s is part of an important shift of attention in science studies. Following research that made a decisive break with Mertonian concerns about science as an institution (see Barnes 1974; Bloor 1976), much of the early laboratory studies take as their primary focus what Knorr Cetina refers to as the "hard core" of science and technology—"its technical content and the production of knowledge" (Knorr Cetina 1995, 140). This work closely examines the processes through which knowledge is itself constituted, and the construction of knowledge as it unfolds in laboratories. Thus, the central method utilized in laboratory studies is broadly ethnographic, though, to varying degrees, the first lab studies used an array of discourse analytic techniques as well.[7]

Collectively, these studies offered several significant conclusions. First, the work points to the importance of the local rather than the universal in knowledge production. Second, this research stresses the centrality of contingent factors rather than the strictly logical or rational in knowledge construction. Third, according to these investigations, laboratory practice is about the creation of order from disorder and the transformation of nature in the laboratory. That is to say, science is not about the simple reading of nature. Indeed, nature itself is transformed in laboratory contexts. Finally, and most generally, this scholarship led researchers to the conclusion that science is, as Knorr Cetina (1981) noted, a process of construction, rather than description.

If it is possible to make these generalizations about the conclusions of the seminal laboratory studies, there is also variation among them; some of the important issues raised by this research have been largely overlooked as this work has become canonical. In the remainder of this section, I describe four of the classic laboratory studies—work by Latour and Woolgar, Lynch, Knorr Cetina, and Traweek—and one more recent contribution by Knorr Cetina to this vital literature. I discuss the major conclusions of these investigations as well as some of the less emphasized aspects of this work. I also provide my assessment of the shortcomings of this scholarship. As should be clear, this book itself is a kind of laboratory study and by providing my perspective on the existing scholarship, I hope to clarify the analytic framework I use in the subsequent chapters.

I begin with Bruno Latour and Steve Woolgar's *Laboratory Life* (1986 [1979]), the first published, and perhaps the most prominent, of the laboratory studies. In 1975, Bruno Latour began an ethnographic study of the neuroendocrinology laboratory of Roger Guillemin at the Salk Institute in California, and although Latour and Woolgar's text covers a great deal of empirical and conceptual ground, of special importance is Latour's account of the struggle in which Guillemin's lab was engaged to define the structure of a hormone now known as Thyrotropin Releasing Factor (TRF).

In assessing Latour's experience in the lab, Latour and Woolgar conclude that scientists are fundamentally writers and readers

(1979, 53), principally engaged in efforts to persuade colleagues of the value and validity of their findings. Laboratory work begins with what Latour and Woolgar refer to as "inscription devices." These are pieces of research equipment, or a combination of them, that transform a material substance into a figure or diagram (1979, 51). According to Latour and Woolgar,

> One important feature of the use of inscription devices in the laboratory is that once the end product, an inscription, is available, all the intermediary steps which made its production possible are forgotten. The diagram or sheet of figures becomes the focus of discussion between participants, and the material processes which gave rise to it are either forgotten or taken for granted as being merely technical matters. A first consequence of the relegation of material processes to the realm of the merely technical is that inscriptions are seen as direct indicators of the substance under study. (1986 [1979], 63)

In the laboratory and in their writing, scientists work to generate and gain the acceptance of particular types of statements (1979, 76ff). The figures and diagrams produced by inscription devices are a central part of this process. According to Latour and Woolgar, scientists work to eliminate the qualifications that are part of statements they make. They seek to "persuade colleagues that they should drop all modalities used in relation to a particular assertion" (1986 [1979], 81). Thus, researchers hope to transform the statement "substance x is believed to do y" to "substance x does y." It is in this context that Latour and Woolgar argue that a fact only comes into being when all temporal qualifications are eliminated[8] and it becomes part of a body of knowledge drawn on by others (1986 [1979], 106). It is in this sense that Latour and Woolgar contend that facts are "constructed."[9]

It seems safe to say that Latour and Woolgar's focus on the construction of facts—the role of inscription devices and the struggle over modalities—is the component of their work that has drawn the most attention and provided them with the largest number of followers. Still, this is not all there is to *Laboratory Life*. Latour and Woolgar devote a chapter to their idea of "cycles of credit." They use this concept to explain the operation of what earlier scholars termed "the scientific community" (Hagstrom 1966). According to Latour and Woolgar, scientists engage in an "endless cycle of investment and conversion" (1986 [1979], 200). They transform money into data and data into prestige. Similarly, arguments can be converted into money or published research papers. There is no definitive order or process of conversion, but careers as well as facts are made in this process. As Latour and Woolgar put it:

> It is at best misleading to argue that scientists are engaged, on the one hand, in the rational production of hard science and, on the other, in political calculation of assets and investments. On the contrary, they are strategists, choosing the most opportune moment, engaging in potentially fruitful collaborations, evaluating and grasping opportunities, and rushing to credited information. (1986 [1979], 213)[10]

Like Latour and Woolgar, Karin Knorr Cetina is interested in understanding science as a constructive, rather than descriptive process. Nevertheless, in her book *The Manufacture of Knowledge* (1981), Knorr Cetina's attention is not directed toward the centrality of persuasion in science or even to the role of instruments in transforming objects of study. Rather, she explores science as a process of selections, and insofar as she is interested in scientists as writers, her concern is with the large gap between what scientific writing says about knowledge production and how that process occurs in the laboratory.

Knorr Cetina's fieldwork was undertaken for a year in the mid-1970s in a government-financed research center in Berkeley, California. The organization was relatively large, employing over three hundred people and covering a wide range of scientific fields from chemistry to engineering and economics. Knorr Cetina spent most of her time observing the work of plant protein researchers.

A central claim Knorr Cetina makes in her work is that science is a highly contingent and local activity in which scientists take strategic advantage of local circumstances. A decision to proceed with an experiment in a specific way or with a particular experiment may be determined less by universal criteria of procedure than by the availability of a particular piece of equipment, the need to justify a specific expense, or the need to use budgeted material before the end of a fiscal year (1981, 9). Knorr Cetina uses the term "indexicality" to point to the contingent and local nature of scientific practice. With this term, she aims to make it clear that

> the products of scientific research are fabricated and negotiated by particular agents at a particular time and place; that these products are carried by the particular interests of these agents, and by local rather than universally valid interpretations; and that the scientific actors play on the very limits of the situational location of their action. In short, . . . the products of science . . . are not the outgrowth of some special scientific rationality to be contrasted with the rationality of social interaction. (1981, 33)

If, as Knorr Cetina suggests, science is local, she also contends that scientific reasoning is not linear, as it is presented in textbooks and embodied in hypothesis testing and the "scientific method." Instead, Knorr Cetina claims that scientists reason analogically. Knorr Cetina found that when scientists explained the origins of ideas they considered to be innovative, they pointed to what they perceived as a similarity between "hitherto unrelated problem contexts" (1981, 52). This approach allows investigators to "bring knowledge from familiar, well-known cases to bear upon an unclear, less familiar, problematic situation" (1981, 57).

Knorr Cetina argues that the local and highly contingent character of scientific practice and the non-linear nature of scientific reasoning that she observed in the laboratory is masked by the form of the typical scientific paper. Knorr Cetina contends that these research products purport to "report" research, but through analyzing paper drafts and contrasting these with laboratory practice, she finds that the history of production is lost; the paper is broken into discrete sections: introduction, methods, data, analysis. In actual practice, these kinds of distinctions "are hopelessly intermingled." The scientists Knorr Cetina observed did not "first perform experiments, then obtain results and finally interpret the outcome" (1981, 121). At some level, the organization of scientific papers obscures and mystifies the process through which results are produced. Knorr Cetina argues that "what is obtained is not independent of *how* it is obtain[ed]" (1981, 122). Through the presentation of discrete sections of methods and results, research papers deny the interdependence of these two aspects of the research process. Finally, given the local and contingent nature of scientific practice, methods sections of published papers are often inadequate for reproduction of research. As Knorr Cetina puts it: "For every bit of published 'method,' there seems to be a bit of unpublished know-how which not only reconstructs the recipe sequence of steps in the paper into that of feasible doings within the situational logic of laboratory action, but also provides routines for diagnosing and coping with many unspecified problems" (1981, 128).

Like Latour and Woolgar, Knorr Cetina directs most of her attention to the laboratory and also spends some time considering the relationship of scientists or labs to the world beyond them. In Knorr Cetina's case, "transscientific fields" is her bridging concept. She argues that "The contextuality observed in the laboratory is constantly traversed and sustained by relationships that transcend the site of research" (1981, 68). She thinks about these relationships variously as symbolic (1981, 82), resource-oriented (1981, 83), and relations of dependency (1981, 77). In Knorr Cetina's framework, a transscientific field may include university administrators, funding agency officials, journal editors, and government functionaries, among others; the configuration of a field will depend upon the issues at stake. Here, participants engage in ongoing struggles over resources and what counts as a resource. According to Knorr Cetina, "resource relationships are at stake, for example, when a position is to be filled by a scientist, when money is to be distributed among scientists or groups of researchers, when a speaker is to be chosen for a scientific lecture, or when a result produced by a scientist is incorporated into the research of others" (1981, 83).

If it is fair to say that the general orientation of most of the early laboratory studies is focused on the laboratory in relative isolation, Latour and Woolgar do extend their analysis briefly to a larger scientific field with their concept of cycles of credit; Knorr Cetina does so with her concept of transscientific fields. Michael Lynch's work, on the other hand, is strictly focused on the laboratory. Indeed, although he does express some interest in scientists' writings, his primary focus is on *talk* and practice in the laboratory itself.

The focus of Lynch's ethnomethodological investigation (1985) is a university brain-sciences laboratory that at the time of his study was working on regenerative processes of the brain. The specific area of the lab's research to which Lynch directed his attention was the electron microscope documentation of "axon sprouting," a type of neural regeneration. What Lynch documents is a radical disjunction between the "unproblematic working out of methodological details" (1985, 4), as these appear in published scientific writing, and what happens in the laboratory. There, reasons and justifications are "suited to the situation at hand and not a transcendental context of inquiry" (1985, xv). Thus, for example, what counts as an artifact in an experimental outcome is likely to be determined in the context of local matters and resolved through laboratory discussion of the local situation. If an experiment fails to produce the expected outcome, scientists ask if it was done correctly or if anything else might have been done to make the experiment work. According to Lynch, such questions arise in the absence of a resolution based on an external standard and are resolved "in a way that is reflexive to the work itself" (1985, 114). Indeed, many of the procedures Lynch confronted in the lab he studied "were not rationalized with available biochemical principles . . . , but were employed because they 'worked'" (1985, 108).

While Latour and Woolgar are interested in the acceptance of the claims of one lab by another, Lynch looks at the ways in which agreement is a "local achievement" produced through talk (1985, 184–188). Following in the tradition of conversation analysis, Lynch closely evaluates sequences of talk. When electron microscope inspections produced unexpected data, Lynch found that no formal criteria existed to measure competent analysis. Instead, "accounts of what was going on were practically settled in members' shop talk" (1985, 191). When there were disagreements about experimental results, Lynch found that "expressions of disagreement" by some lab members resulted in other members shifting their positions (1985, 202). According to Lynch, when scientists do not have "independent or privileged access to 'objective features' of objects," they engaged in redescriptions in pursuit of agreement (1985, 202). Often this involved enveloping the claims of another person into one's own statement. Lynch also found that questions of standards—"how much," "how good," "how frequent"—are often resolved for practical purposes. An assay, for example, might be judged to be "good enough." As Lynch puts it: "We find matters of 'how much is enough' solved for all practical purposes by it being asserted that something is 'clean,' if not 'sterile,' or 'very few' rather than 'too many'" (1985, 262).

Sharon Traweek's laboratory study, *Beamtimes and Life Times* (1988), is of a different variety altogether.[11] First, while laboratories in the biological sciences are the subject of the work of Latour and Woolgar, Knorr Cetina, and Lynch, Traweek studies particle physicists. Her research is centrally focused on two laboratories: the Stanford Linear Accelerators (SLAC) near San Francisco and the National Laboratory for High Energy Physics (KEK) near Tsukuba, Japan. However, because high energy physics depends on large and expensive instruments, her work is really about the community or national communities of particle physicists. Furthermore, while these three other laboratory ethnographies explore fact-making and knowledge production, Traweek provides "an account of how high energy physicists see their own world; how they have forged a research community for themselves, how they turn novices into physicists, and how their community works to produce knowledge" (1988, 1). Thus, her book traverses the social organization, developmental cycle, cosmology, and material culture of high energy physicists.

Traweek makes far fewer general comments about science and scientists than do other authors of laboratory studies. This may be due in part to the implicit comparison she makes between biology and physics. According to Traweek, part of the distinctiveness of high energy physics is the technology that it requires:

> The place of detectors in high energy physics contrasts sharply with the role of research equipment in many other fields. In much biological research, for example, laboratory machines are not built or even designed by the scientists using them; they are mass-produced and advertised in catalogs. Their design incorporates theories and laboratory practices so widely accepted that their validity has not been questioned for many years, perhaps decades. (1988, 49)

From the early work of Latour and Woolgar (1979) to the later work of Joan Fujimura (1996), a great deal of attention is paid to the creation of black-box technology in biological science and the ways in which the science underlying these tools comes to be taken for granted. Researchers utilize them, implicitly agreeing on their appropriateness. In contrast, we learn from Traweek that high energy physics detectors are not black boxes. Indeed, in high energy physics "inventing machines is part of discovering nature" (1988, 49).

Beyond an implicit comparison with biology and the evidence this gives her regarding the distinctiveness of physics, Traweek's cross-national comparison provides her with insight into the role of national culture in shaping research institutions and practices—how it creates distinctive national styles in physics. For example, differences between funding procedures in the United States and Japan are associated with different particle detector construction practices, and these practices are, in turn, associated with different approaches to experimentation. At SLAC, detectors can be modified as a result of collaborations between engineers and experimentalists. This allows for trail and error adjustments in detector design (1988, 148). By contrast, in Japan large initial outlays for detectors are provided by the government for detector development, but little or nothing is provided for subsequent modification. Consequently, according to Traweek, "Japanese physicists were inclined to design all-purpose durable detectors and to push for the most sophisticated technology available for each component" (1988, 70). KEK physicists design detectors in collaboration with engineers from private firms. A firm then constructs the detector on site. Because there is little money available for later modification, there is no point in having workshops and technicians on site to engage in a modification process.

Unlike any of the other key laboratory study texts, Traweek's analysis examines the role of gender. Traweek provides data on the division of labor by gender in physics in both the United States and Japan. She explore the role of stereotypical images of women and gender relations in conversation and textbooks and how these images shape attitudes among students, especially attitudes about who can and should be a physicist. In this context, a successful physicist is a married man and has an understanding wife with no independent career goals (1988, 83, 84). Furthermore, the characteristics associated with success are those associated with maleness in western culture: independence, experience, competitiveness, and individual orientation (1988, 104).

Before I turn to my evaluation of these laboratory studies and consider how my project is linked to them, I want to briefly discuss a relatively new major laboratory ethnography, Knorr Cetina's *Epistemic Cultures* (1999). This study marks a considerable move from earlier explorations of laboratory life. Here, Knorr Cetina's attention is not trained on knowledge production, but on the cultural and organizational environments where knowledge is produced. In this context, Knorr Cetina understands epistemic cultures as "cultures that create and warrant knowledge" (1999, 1). And there is not just one epistemic culture. Indeed, this work breaks with several of the early analyses of laboratories and much other work in science studies in that Knorr Cetina rejects the idea that there is one science. Instead, drawing on ethnographic data from several sites, she seeks to understand differences between high energy physics and molecular biology (1999, 4, 22).

The objects that are studied by molecular biologists and high energy physicists are distinctive in Knorr Cetina's view. In the former case, the phenomena of study can be captured and manipulated. By contrast, the objects studied in high energy physics "are too small ever to be seen except indirectly, too fast to be captured and contained in a laboratory space, and too dangerous as particle beams to be handled directly" (1999, 48). But perhaps more interesting is the different organizational characteristics between these two fields. In high energy physics, research tools are extraordinarily expensive, technologically complex, and often quite large. These factors demand that large scale collaborations be formed (1999, 114). The collaborative environments in which high energy physicists work are characterized by Knorr Cetina as "post-traditional communitarian structures." She describes these as "structural forms attempting to implement collective ways of working that downgrade the individual as an epistemic subject and that emphasize instead such communitarian mechanisms as collective ownership and 'free' circulation of work" (1999, 165; see also 164). Experiments are made up of legally and financially independent institutes, not individuals, and hierarchy is replaced by "horizontal links between scientists, or groups of scientists, and objects" (1999, 172). Since individual epistemic subjects do not exist in high energy physics, there is a great incentive to cooperate (1999, 171). Finally, because of this context, authorship can be characterized as collective. Although all members of a collaboration are listed on papers, papers are associated with experiments, and so the experiment has "epistemic agency." It is, according to Knorr Cetina, "the work of the experiment that travels through the community under the experiment's name, not scientists together with their work" (1999, 168).

By contrast, in the molecular biology laboratories Knorr Cetina studied she found the individual is the epistemic subject (1999, 217). The individual nature of the endeavor is revealed in the fact that members of a laboratory have their own projects. Consequently, authorship is claimed by individuals (1999, 217, 220). Even the molecular biology laboratory, which might be regarded as a collective entity, is organized around an individual: the laboratory leader. In this context, tensions can emerge because the lab and the leader require that only some projects be successful, but the individual's career advancement demands success on his or her particular project (1999, 229–231).

There can be little doubt that laboratory studies have played a substantial role in making science studies a vital field of inquiry and continue to do so. I place my work squarely in that tradition of laboratory ethnography. But while it is probably fair to categorize the work of Latour and Woolgar, Knorr Cetina's *Manufacture of Knowledge,* and Lynch's first book as "ethnographies of knowledge," my own study is better described by the rubric, "ethnography of scientific practice." My interest in practice, as I have made clear, is in how the broad work in the laboratory is shaped by the larger environment in which university laboratories are embedded. In this way my study differs considerably from those studies cited above.[12] Although Latour and Woolgar develop their concept of "cycles of credit," and Knorr Cetina works with the notion of "transscientific fields," the book-length works by Latour and Woolgar, Knorr Cetina, and Lynch all focus on laboratories in relative isolation.[13]

Even to the extent that Latour and Woolgar and Knorr Cetina do develop concepts that point beyond the laboratory, in my view, they are inadequate. Knorr Cetina rightly points out that Latour and Woolgar's treatment of science as a market ultimately leads to the view of a "self-contained quasi-independent system" (1981, 73). Such an idea cannot easily accommodate consideration of the relationship between scientific practice and the world of commerce. Furthermore, even when it seems that consideration of larger structural features of the environment is appropriate, Latour and Woolgar sometimes turn away.[14] For example, at one point, their use of the passive voice in a sentence suggests something is missing. According to Latour and Woolgar, "since Guillemin wanted to determine the *sequence* of TRF, and since he was ready to reshape a subfield around this crucial goal, new standards were set as to what could and could not be judged reliable" (1979, 121). But who set these standards? Why were they able to do so? Why were the standards accepted by relevant scientists? Did not the larger environment in which Guillemin's lab was embedded have any influence on his success in redefining the field?

Knorr Cetina recognizes the importance organizations and macro-level structures can have in shaping social action, and I believe she is correct; but she also contends that much about social action needs to be explained at the level of interaction (1981, 44). I do not dispute this, but if we focus primarily on interactions, we may overlook larger factors that influence the interactions. In this context, Lynch never explains why interactions end in agreement. Does the social authority granted by structural location affect whether subordinate actors concede?[15] How does disciplinary socialization affect the resolution of the negotiations with which Lynch is concerned?

Knorr Cetina admits that rules can shape action, but she focuses her attention on the ways in which people involved in a negotiation process actively manipulate the rules (1981, 45, 46). Of course, this is an important angle for investigation, but when manipulation of rules and the use of rules as resources displaces the constraining role of rules as an object of investigation, an important component of social analysis is lost. Certainly, rules defining play within the intellectual property regime, for example, are not as easily manipulated by all participants. Indeed, in the case of university scientists they are often constraining. Finally, Knorr Cetina is attentive to the role of power in practice of science—a crucial factor in understanding the ways in which the world of commerce affects the practice of university biology. However, her description of power as "a symmetrical, albeit unbalanced relationship" stresses the space for scientist agency and, I fear, underplays the importance of structural constraints to scientist action (1981, 46).

In principle, I wonder if one can ever really understand laboratory life by treating laboratories in isolation. More to the point, however, when we investigate the intimate connection between university biological sciences and the world of commerce, I believe it is impossible to fully understand daily practices and epistemic cultures without acknowledging and empirically documenting these links. It is particularly surprising to me that while both Latour and Woolgar (1986 [1979]) and Knorr Cetina (1981) talk about the research done at the laboratories they studied as straddling the boundary of basic and applied investigation, they do not seriously explore the relationship between their labs and the commercial world. Indeed, the factors that concern me in the chapters to come, particularly intellectual property considerations and commercial production of research tools, are sometimes mentioned in passing by these authors, but not analyzed.

My own broadly structural framework—which I outline later in this chapter—places power and resource asymmetry at the center of analysis.[16] Perhaps such a framework is less necessary when laboratories are treated in isolation, but again it seems to me that the role of these features in shaping laboratory life has been underplayed or underdeveloped in early book-length ethnographies of knowledge (or, as in Lynch's case, entirely ignored). By contrast, in quite different ways, Traweek and Knorr Cetina (1999) both systematically link the laboratory to some larger environment and provide analysis with organizational or structural elements. It seems to me that Traweek and Knorr Cetina (though to a lesser extent) are appropriately attentive to the ways in which funding regimes and patterns affect the organization of science and ultimately experimental design. On the other hand, it is surprising that Knorr Cetina, writing in 1999, gives such little attention to intellectual property considerations in molecular biology, even as she is attentive to the social organization of authorship in the field. More generally, for molecular biology (though less relevant perhaps for high energy physics), if we understand the term "epistemic cultures" to refer to "the different practices of creating and warranting knowledge in different domains" (Knorr Cetina 1999, 246), it seems appropriate that the relationship between laboratory life and the world of commerce receive more attention.

Actor Networks and Social Worlds

In this final section of the chapter, I provide an analysis of two agency-centered approaches—actor-network and social-worlds theory[17]—to the study of technoscience that have held central positions in debates in the social studies of science over the past several years.[18] I use this scholarship as an entrée into a brief elaboration of the framework that guides my analysis in the chapters to follow.

Actors—human and non-human—and boundaries, distinctions, and dichotomies are central to the actor-network approach. We are advised to follow the actors wherever they lead and to see the world from their perspective. From this starting point, advocates of this system focus their attention on construction or building, and it is deemed inappropriate to make any a priori assumptions about context and content, inside and outside, science and society, arenas and actors (see Latour 1987). Similarly, the identities and interests of the actors are seen as products of constituting processes (see Cambrosio, Limoges, and Pronovost 1991; Wynne 1992). For the most part, proponents of this approach present agent-centric analyses in which actors operate in highly manipulable worlds that are subject to minimal constraint (Moore 1995). Although the focus tends to be on actors' constitutive activities—their active building roles—analysts stress that causality (although they probably would not want to use the term) is not unidirectional. Instead, either the world is a seamless web (Shapin 1988) in which the evaluation of the direction of causes and effects is impossible, or, as in Law and Callon's later work, elements of the world are *mutually shaping*, particularly networks and actors (Law and Callon 1992, 25). Finally, enrollment is a central concept in actor-network analyses. Perhaps the best way to enroll others is to offer something that helps them reach their goal in such a way that their actions will also advance your goal. Enrollment, then, involves the translation of the other's interest into one's own terms; in the best case, one makes oneself "indispensable" to the efforts of others (Latour 1987).

Callon and Law have modified and clarified this framework in recent years. It is too simple, they claim, "to say that context influences, and is simultaneously influenced by, content." They propose the alternative concepts of "global network" and "local network." A global network, as Law and Callon define it,

> is a set of relations between an actor and its neighbors on the one hand, and between those neighbors on the other. It is a network that is built up, deliberately or otherwise, and that generates a space, a period of time, and a set of resources in which innovation may take place. Within this space . . . the process of building a project may be treated as the elaboration of a *local network*—that is, the development of an array of the heterogeneous set of bits and pieces that is necessary to the successful production of any working device. (1992, 21, 22; emphasis added)

Law and Callon argue further that the ideas of content and context may be transcended "if projects are treated as *balancing acts* in which heterogeneous elements from both "inside" and "outside" the project are juxtaposed" (1992, 22; emphasis added).

Law has devoted greater attention to conceptualizing power than early actor-network researchers did in their studies. According to Law, power is a function of the network of relations in which an actor is implicated. From this perspective, power can be understood as a condition or a set of conditions, but must also be understood as an effect. Moreover, power is a product of "more or less precariously structured relations" (1991b, 170) and is itself precarious (Law 1991b, 177).

The social-worlds approach has much in common with actor-network theory.[19] Like work in the actor-network tradition, the studies of social-worlds proponents are frequently agency-centered and emphasize construction. For example, in her 1996 book, *Crafting Science*, Fujimura describes her work as an exploration of the "representational, organizational, and rhetorical work done by researchers, students, sponsors, and audiences *to create* the 'world' of proto-oncogene research" (1996, 2; emphasis added). In addition, in her analyses she does not consider unidirectional causes, instead she stresses interactions between different worlds and "co-construction" (Fujimura 1996, 11).

Like scholars employing actor-network approaches, social-worlds analysts advise us against creating separations such as context and content, outside and inside. Clark and Fujimura (1992) suggest a kind of indivisibility of the realm of the laboratory and the other "worlds" with which it is connected. They call on us to think about all of the factors outside of the lab as part of the "situation itself" rather than as "context." In many instances, "the world is in the laboratory and the laboratory is in the world," as Fujimura puts it (Fujimura 1996, 11).

Like some work in the actor-network tradition, emphasis in the social-worlds approach is placed on the *mutual benefits* of relations. Analysts focus attention on the ways in which "standardized packages" (Fujimura 1987, 1988, 1992, 1996) and "boundary objects" (Star and Griesemer 1989) make cooperation between social worlds possible, not on the ways in which some worlds may shape the practices of others.

In her 1996 book, Fujimura explicitly conceptualizes power. She suggests that power and authority are "distributed among different actors, objects, and social worlds" (1996, 17). Citing Antonio Gramsci, Fujimura declares further that "the process of standardizing technologies is a process of establishing *hegemony* . . . over other ways of knowing" (1996, 72). According to Fujimura, hegemony and power are attributes of technologies, "paradigms, and ultimately standardized packages" (1992, 153, 114).

Although both actor-network and social-worlds research have enlivened the field and appropriately cautioned against reductionism, both suffer from several important and related methodological and conceptual shortcomings. First, emphasis on agency has led analysts to ignore or minimize the *constraints* placed on agents in their efforts to act.[20] At a methodological level, restricting analysis to the world as it is perceived by actors may, as Law notes, lead us to ignore "distributions [of resources, for example,] . . . that are of no concern to the actor being followed" (1991a, 11). Furthermore, we may, overlook institutional constraints which actors may not notice.

A second limitation of actor-network and social-worlds approaches is their virtually exclusive emphasis on processes of construction. Active agents are always constructing their worlds in this work. But emphasis on construction turns our gaze away from the effects of the already existing attributes of the world in which science is practiced. Furthermore, focus on co-construction, interactions, and dialectical relations leads to the implication that it is analytically impossible to isolate unidirectional causal relations. Analysts may thereby fail to investigate structural effects.

Related to the focus on construction is the desire within these approaches to transcend dichotomies. In my view, it is not immediately obvious why we should need to "transcend" the context/content dichotomy. Indeed, it seems to me there may be times when looking at the effects of the world "outside" the lab on the practices "inside" is worthwhile. Furthermore, while we need not deny that projects or practices can result from "balancing acts" and juxtapositions, I see no reason to assume this either. Indeed, doing so might very well lead us to overlook cases in which there are asymmetries in the efficacy of factors involved in the construction of projects or shaping of practices.

A final and related weakness in the actor-network and social-worlds traditions is rooted in their perception of power.[21] The mutuality of relations is stressed in both approaches. However, this orientation underemphasizes situations in which the benefits of relations may be asymmetrical. In the actor-network approach, translation implies a kind of ultimate compatibility of interests in which benefits of alliance are mutual. Such a portrait may lead us to ignore the possibility that enrolled actors may benefit less than the enrollers and that without the actions of enrollers, the enrolled might have acted differently. Indeed, careful work in the history of science and technology (see Kloppenburg 1988; Leslie 1993; Noble 1984) shows that power inequalities affect outcomes and that not all relations in the world of technoscience are mutually beneficial.

I find Fujimura's use of the term "hegemony" in her conceptualization of power particularly inadequate. Gramsci, of course, defined the term in many different ways. Common usage draws attention to the notion that domination may occur with the consent of subordinate actors. This may happen because the realization of "real" interests of subordinate groups becomes inextricably linked to the realization of dominant group interests, or the world view of the dominant group may become the world view of society as a whole. Indeed, central to Gramsci's formulation is a view of society in which relations of domination and subordination are a pivotal feature (see Mouffe 1979). However, such relationships are largely absent from Fujimura's conception of the term.

In my view, Law's work from the early 1990s marks an appropriate recasting of the concept of power in science and technology studies, although his emphasis still diverts attention from the *effects* of power and inequality. Stressing the precarious character of power relations, Law expresses his concern with *"how it is that relations are stabilized for long enough to generate the effects and so the conditions of power"* (1991b, 172). In other words, Law is interested in the *constitution* or *construction* of power, not with its *effects.* Finally, I am not convinced that the assumption that power relations are always precarious is an appropriate position with which to begin an analysis. Historically, such relations seem reasonably stable over time.[22]

In this book, I employ a broadly structural, organizational and/or institutional perspective,[23] and I use these terms interchangeably. However they are named, these entities can be seen to constitute specific formal and informal, explicit and implicit "rules of play." These rules of play establish distinctive resource distributions, capacities and incapacities, and define specific constraints and opportunities for actors depending upon structural location (Lindberg 1982). It is within this context that I understand power and its operation. The rules of play that define structures convey advantages to certain actors over others by endowing them with valued resources—or even by serving as resources themselves. In the later case, enforcement of rules may benefit certain actors to the detriment of others. Here, power can be said to exist in at least two senses. In the simplest behaviorist approach, when the structural location of one actor enables it to shape the action(s) of another actor in ways in which the latter might not otherwise act, one can say that power has been exercised. In the more complicated case, as Lukes described it, "the bias of the system is not sustained simply by a series of individually chosen acts, but also, most importantly by the socially structured and culturally patterned behaviour of groups and practices of institutions" (1974, 21, 22). Such bias can "unevenly [distribute] influence, access, and control" (Wright 1994, 11). In the first instance, agents can be understood to have power, although the basis of this power is organizational. In the second instance, power is a direct attribute of structures and need not be enacted to have effects.

Advocates of actor-network theory and the social-worlds approach might assert that the framework I present here reifies structures, portraying "these as pre-existing, fixed and sovereign—rather than themselves only partially formed, indeterminate and open to construction through processes of negotiation" (Wynne 1992, 577). Ironically, for analysts who wish to avoid distinctions, this kind of response sets up a dichotomy: structures, actors, interests, and identities must be viewed as either always in the process of construction and reconstruction, or analysts accept them as already existing when actors arrive on the scene. In the latter case, analysts are accused of reifying the world they are studying. But this is not an either/or issue; there is a possible third position: structures, interests, actors, and identities are constructed historically, but at any given time, these phenomena have an established character, and this configuration has effects. In the case of structures, human and organizational actors are likely to confront them as "external" and sometimes "constraining."[24]

Indeed, I do not deny that these structures are constructed. Only that at any given time actors will confront structures that are already constructed, and that these entities will sometimes shape practices. If one must always focus analysis on the construction process, then one can never explore the influence or effects of already constituted structures. Furthermore, if we assume that entities like "content" and "context" or the "micro" and the "macro" exist in dialectical relationship at every instant—if they are always in the process of co-construction—then we cannot explore how one might shape the other. I do not dispute that such relationships can be reciprocal over time, and I do not wish to present a world in which certain actors *always* lack agency. However, in this study I take a synchronic look at one laboratory; I do not focus on how the laboratory shapes the environment in which it is embedded, but rather I focus centrally on how the environment shapes it.

If structures, although already constructed, have histories, then it is legitimate to begin one's analysis by assuming the existence of particular structures. This might be viewed as a pragmatic move (Wynne 1992, 579), as it offers a starting place and enables one to initiate investigation. That is the approach I take here. I entered the laboratory that is the focus of my investigation knowing that the lab had relationships with university administrative entities, with for-profit companies with which the laboratory collaborates, and with commercial suppliers of research materials. I knew, furthermore, that matters of intellectual property concerned the lab leader and some lab members. I did not know what these factors meant for lab practices or how they might affect laboratory life. Investigating this is a central aim of my project.

Throughout this book, I assume that university laboratories are situated within a complex institutional matrix and that although all aspects of this configuration are subject to change, at any given point these structures are stable and can influence laboratory practice. Given this assumption, my approach might be viewed as an ethnography in which my attention is focused on what Michael Burawoy refers to as "the 'macro' determinations of everyday life" (1991b, 271). Like Burawoy, I am interested in understanding the ways in which "the situation is shaped from above rather than constructed from below" (1991b, 276). I am attentive to the specificity of the laboratory I am studying, but I am more generally concerned with what this case can tell us about the world in which a particular lab and many other university laboratories are embedded.[25]

Conclusion

In this chapter, I have covered a wide conceptual terrain. My aim in doing so was to clearly and explicitly place my own orientation in relationship to scholarship that I find relevant. My approach in the pages that follow can be summarized quite briefly: I undertake an analysis that, contrary to much work on UIRs, stresses the *indirect* effects of the world of commerce on the daily practices of university biology. To do this, I move beyond the micro-level focus of many of the early ethnographies of knowledge, and attempt to link the micro and macro (concepts that I realize are currently out of favor). Finally, my analysis emphasizes how the larger world constrains university biology laboratories that are embedded in that world, and may work against opportunities for agency and cooperation. Despite my criticisms of the literatures I discuss above, I have found all of them intellectually stimulating, and if my work prompts debate within any or all of these research areas I will be both honored and pleased.

3

Braided Paths

The Intertwined Development of Biocontrol Research and Agro-Industry

1882. 1889. 1902. 1945. 1983. 1985. These dates mark pivotal points in the braided history of chemical treatment of agricultural plant pests and of an alternative means of protecting plants from disease and insects: biocontrol. The first chemical fungicide was discovered in 1882. The first instance of an insect outbreak being controlled through the use of another insect occurred in 1889. In 1902, an early entrant into the agri-chemical industry began to manufacture and formulate chemicals for farm use. The years around 1945 heralded the modern age of agrichemicals with the development of organic compounds highly effective against insects and weeds. In 1983, metalaxyl—the fungicide with which UW85 may eventually compete—was approved for agricultural use by the U.S. government; and in 1985, workers in the Handelsman lab determined that a *Bacillus cereus* strain they called UW85 had biocontrol properties.

These five discoveries over about a century reveal the intimate connection between the history of agrichemicals and the history of biocontrol. I will argue that it is impossible to understand the history of biocontrol—and thus of UW85—without scrutinizing its relationship to the development of agrichemicals. Thus, my aim in this chapter is to trace and examine this intertwined history. My objective here is to illustrate one way in which the world beyond a university laboratory can shape the practices of that laboratory, as well as an entire field of scientific investigation. My basic claim is that the dominant role taken by the agrichemical industry in defining pest-control strategies throughout the twentieth century, but especially in the years after World War II, had and continues to have significant effects on the scientific and scholarly practices of workers engaged biocontrol research, including members of the Handelsman lab. The issues of structure and power that I considered in chapter 2 are central here. This discussion illustrates an instance in which power is exerted directly at the outset: the research that was done was determined by the kind of research that was financially supported, and this research could only be promoted by organizations with financial clout. After research in the area of agrichemicals became a commercial success, it is possible to see the effects of the agrichemical industry on biocontrol research as pervasive, but *indirect*. Here, power is not exercised directly by individuals or even by organizations. Instead, a set of guidelines for action are shaped by a structure that became established over time. These "rules" constrain the behavior of actors and do not depend on the direct intervention of powerful individuals to have this effect.

Chapter 3 is organized this way: First, I discuss early pest problems and pest control in agriculture. Next, I examine the pivotal role of World War II in shaping pest-control strategies in the postwar period. I follow this discussion with an historical examination of biocontrol and its relationship to agrichemicals. My discussion of biocontrol is capped by two case studies: an exploration of insect control in the citrus industry and an analysis of the development of UW85.

Pest Problems and Pest Control: The Early History

The folklore of chemical control of fungi that cause plant disease suggests a literary source as the first written discussion of a chemical fungicide. Homer's *Iliad* and *Odyssey* are cited by several sources as texts that mention the fungicidal properties of sulfur (Carley 1994, C18; Hayley 1994, C2; McNew 1959, 42). Still, it was not until the late nineteenth century that particular chemical compounds are mentioned as pest-control agents. Prior to the

1870s, "cultural and physical control" practices, including crop rotation, destruction of crop refuse, timing of planting to avoid periods of high pest density, and physical separation of crops, were the techniques commonly used to control plant pests (Osteen and Szmedra 1989, 5). According to one analyst, the 1868 discovery by an unknown inventor that the dye known as Paris green "could kill insects launched a new era in the deliberate use of toxic substances for commercial purposes" (Perkins 1982, 3). Other substances, including sulfur, of course, were known and used for their pesticidal properties prior to 1868, but only one other substance was traded and produced on a significant scale before Paris green. Furthermore, Paris green was the first agricultural pesticide other than sulfur for which the "toxic principle" was known. The toxin was arsenic, a substance later used in other pesticides. According to John Perkins, by 1910 Paris green and lead arsenate—a pesticide developed after Paris green—were the most widely used commercial insecticides, with annual sales estimated at $20 million (1982, 3).

In 1882, Bordeaux mixture was developed; it is considered by many to be the first agricultural fungicide (Carley 1994, C19; McNew 1959, 42). The need for this fungicide grew out of the grape downy mildew infection of vast vineyard acreage in France. The disease apparently came from the United States around 1870. The discovery of Bordeaux mixture's efficacy was quite accidental. It was observed that a mixture of lime and copper sulfate splashed on vines to discourage theft appeared to control downy mildew. After several years of experimentation, Millardet claimed discovery of Bordeaux mixture, a fungicide composed of copper sulfate, calcium oxide and water (Carley 1994, C19). This copper sulfate-based substance was introduced into the United States and England soon afterward and was used successfully to control the destruction of potatoes by *Phytophthora infestans*. By World War II, twenty million pounds of copper sulfate were being used annually to prevent disease of potatoes, and nearly as much was used to control diseases that attacked apples (McNew 1959, 43).

The two World Wars mark decisive points in the development of agricultural chemicals. World War I promoted the development of insecticides based on synthetic organic compounds. In 1914, when war broke out in Europe, the United

States had virtually no ability to produce coal-tar intermediaries, "the lifeblood of the synthetic chemicals industry" (Perkins 1982, 5). Before the War, the United States relied on German-produced chemicals to support its limited dye industry. When hostilities ended trade relations between the two countries, chemical producers in the United States initiated manufacture of domestic coal-tar intermediaries, and these provided the basis for dye, medicine, chemical and explosives manufacture (Perkins 1982, 5). A pivotal wartime chemical, Chloropicrin, was developed initially as a tear gas and was discovered to have insecticidal properties in 1916. Sometime later, Chloropicrin was also found to be effective against soil borne diseases (Perkins 1982, 5; Carley 1994, C19). Paradichlorobenzene, a byproduct of chemicals used in explosives during World War I, was later determined to be highly effective against the peach tree borer. More generally, according to one analyst, "the manufacture of explosives had a direct spin-off effect on the development of insecticides that presaged the later successes of DDT" after World War II (Perkins 1982, 5).

If World War I stimulated important developments in chemical agriculture, it was World War II that decisively shaped modern agricultural pest control and the future of biocontrol. The need for increased food production and insect control for military purposes prompted the U.S. government to promote and coordinate pesticide development and production. According to Palladino, "Since the main preoccupation of the military was to protect troops always on the move, it was far easier to douse these troops with repellents, or to spray their temporary quarters, than to introduce natural enemies or drain any infested swamps through which they moved. Work on biological and cultural control was much less important, and was de-emphasized even further than it was before" (1996, 37).

The pesticide agent DDT, developed by the Swiss manufacturer Geigy, became a bright star in the burgeoning field of chemical pest control in agriculture.[1] The company's research indicated the substance's usefulness as a delousing agent and as an agricultural chemical (Perkins 1982, 9). But it was the protection DDT offered to American troops against malaria and other insect borne diseases that transformed the chemical, according to Palladino, into a "technological wonder" (1996, 37).

Chemical developments stemming from World War II research fundamentally altered the character of agriculture. The U.S. Department of Agriculture (USDA) estimated that some twenty-five new pesticides were introduced between 1945 and 1953, and sales of insecticides alone climbed from $9 million in 1939 to $170 million in 1954 (Palladino 1996, 37). Research and use of new insecticides altered the practices of farmers, the nature of scientific research, and the character of the chemical industry (Perkins 1982, 11).[2] Synthetic organic chemicals were relatively cheap, effective, and more target-specific than previous chemicals. Combined with developments in plant breeding, new agricultural chemicals increased yields throughout agriculture—even in areas previously not considered suitable for chemical control—by inhibiting weed growth, insects, and fungal pests (Perkins 1982, 11; Harrington 1996, 415; Kohn 1987, 162, 163). From the end of WWII through the early 1960s, for example, corn production per acre nearly doubled from 36.5 bushels to 68.1 bushels per acre. Wheat productivity jumped dramatically too, from 16 to nearly 26 bushels per acre. The same was true for many other crops (Kohn 1987).

One analyst has written: "Growth was a central theme of the American post–World War II economy. More was better. Expansion was good. Using technology to become larger and produce more was part of the shared consensus in the postwar economy" (Harrington 1996, 415). In agriculture, this meant substantial increases in agri-chemical use. By 1950, organic chemical pesticides accounted for 60 percent of the U.S. pesticide market, and this figure reached 90 percent by the mid-fifties (Kohn 1987, 162, 163). Between 1950 and 1960, the agricultural land treated with herbicides to control weeds more than doubled to some 50 million acres, at a cost of almost $137 million. In 1960, farmers and ranchers covered nearly seventy million acres with about 165 million pounds of insecticide products (Harrington 1996, 420). Overall, between 1947 and 1960, synthetic pesticide production increased fivefold; by 1981, it had reached 1.8 billion pounds annually (Doyle 1985, 183).[3] Annual sales in 1984 topped the $5 billion mark (Hayley 1984, 52). In 1991, American farmers spent over $1.3 billion on insecticides and another $4.1 billion on herbicides and fungicides (Palladino 1996, 8).

The Stunted History of Biological Control

It is within the context of the formidable success of agrichemicals, especially in the years following World War II, that work in biological control must be considered. It is possible to examine the impact of historical developments in agrichemicals on biocontrol on two levels. First, as I discuss in this section, there are the global and *direct* effects. Here we can see the impact that the dominance of the chemical industry has had on the character of biological control as a research field and a commercial venture. A substantial scientific basis in understanding agricultural chemicals as well as plentiful resources for new developments meant there was far less attention directed at biological control than there might have been. Biological control, thus, displayed fewer successes and seemed to have limited commercial potential. Second, the chemical industry's dominance of pest-control research *indirectly* affected—and continues to affect—research on biocontrol at several levels, including such varied things as the framing of grant applications and research papers, the design of field tests, the reporting of research results, the selection of tools used in scientific investigation, and in decisions regarding what counts as successful disease suppression.

Biological control or biocontrol simply means the use of biological materials (rather than synthetic organic chemicals) to control natural enemies of plants (Sawyer 1996, xvi). These enemies can be insects, which are often controlled by other insects, or plant diseases, which can be inhibited by bacteria or fungi. The former area is typically the domain of entomologists and the latter of plant pathologists. By the middle of the nineteenth century, leading entomologists in the United States were aware of the foreign origins of "this country's most devastating pests" (Caltagirone and Doutt 1989, 2). These scientists called for the importation of natural enemies to combat these pests. However, although the term "biological control" was first used in 1919 (Cook and Baker 1983, 33), many view the effort to control the citrus pest, cottony cushion scale, in 1889 as the beginning of biological control as a "scientific strategy" (see Morrison 1989, 4; Sawyer 1996, xvii). Cottony cushion scale—known technically by its Latin name, *Icerya purchasi*—was believed to have been introduced to the United States from Australia, possibly in a nursery shipment to the San Francisco Bay Area (Sawyer 1996, 9; Caltagirone and Doutt 1989, 3). Described by one commentator as having little effect in its indigenous Australian ecosystem, in American citrus groves, this quarter-inch-long insect that looked like a small white piece of fluted wax, fed on the sap of fruit trees, thereby weakening them (Sawyer 1996, 9).

A USDA explorer working in Australia determined that the female red and black vedalia beetle *(Rodolia cardinalis)* kept the scale in check in its indigenous environment by feeding on it. With this knowledge and at minimal cost, the USDA explorer collected beetles in Australia for use in American citrus groves. The beetle reproduced, spreading through affected citrus growing areas, and after only one year the beetles had "dramatically reduced the scale population" (Morrison 1989, 4). Importantly, according to one commentator, this development occurred at a time when there were "few chemicals of value for insect suppression" (Greathead 1989, 189).

Biological control of plant diseases came later. One study notes the first attempts were made in the years between 1920 and 1940, and the "first demonstration that a total antagonistic soil microflora could be transferred to produce pathogen-suppressive soil" occurred in 1931 (Cook and Baker 1983, 33, 36). Efforts to control weeds biologically were not attempted until the mid-1940s, when insects were introduced to control Saint John's Wort (Morrison 1989, 5).

Many believe that the development of biological control was stunted in its early history because of the extensive attention paid to chemical controls and their stunning successes following World War II. Writing in the entomology journal *Antenna,* David Greathead notes that "important changes in insect control began in 1946 when DDT became available and, along with other powerful, broad spectrum, persistent organochlorine pesticides, appeared to offer a quick and simple solution to all insect pest problems. Consequently, the need for biological control was questioned" (1994, 192). Morrison makes a related point in his suggestion that "despite biocontrol's successes, in the 1940s chemical insecticides distracted scientific attention from all aspects of biocontrol" (1989, 5). Finally, Campbell notes in a relatively recent plant pathology text that "the success of chemical pesticides in controlling plant diseases (and insect pests) has worked against biological control. While there were very effective, relatively cheap methods available there was no incentive for the development or marketing of other systems" (1989, 44, 45).[4] It is hard to know if biological control "could have reached the same degree of success as chemical control has, given the same amount of research time and finance" (Campbell 1989, 46), but the data that exist indicate a dramatic divergence in the postwar attention accorded to biological and chemical control of plant pests. Evidence from the late 1960s suggests that virtually one hundred times more money was devoted to pesticide research than to biocontrol studies (Ordish 1976, 212). This unequal support is strongly evident in the number of papers presented at conferences. Between 1963 and 1973, the ratio of chemical control to biological control papers presented at the annual meetings of the British Crop Protection Council neared thirty-five to one (Ordish 1976, 213). Similarly, after 1946 the proportion of notices of biological control papers in the *Review of Applied Entomology* fell with the explosion of research published on chemical pesticides (Greathead 1994, 192).[5] This trend is evident in the *Journal of Economic Entomology* as well, where the percentage of papers on "the general biology of insect pests and their biological control properties dropped from 33% in 1937 to 17% in 1947, while articles devoted to the testing of insecticides rose from 58% to 76%" (Perkins 1982, 11). Even as late as the 1990s when the problems with chemical agriculture were well known, fewer than 10 percent of the nearly two thousand scientists affiliated with USDA's Agriculture Research Service were engaged in research on natural crop protection methods, and less than 10 percent of the Service's nearly $500 million research budget supported work on such methods (Groves 1998, D7).

The limited extent of biological control research and the massive amounts of money showered on chemical research undoubtedly affected the types of pest-control techniques ultimately used by farmers. Farmers seek consistent, reliable, cost-effective, and easily applied means of pest control; they "will only use difficult procedures with expensive and unreliable control agents if there is nothing else that is legally available" (Campbell 1989, 47). Biological control has had limited commercial success at least in part because of the limited resources that have been devoted to its study. Control of cottony cushion scale in California citrus orchards was an unambiguous and high-profile victory, but a low level of consistency and reliability plagued much of the work in the decades that followed. In the pro-growth environment of the postwar period (Harrington 1996), farmers who failed to adopt chemical pest control would likely be forced out of business by farmers who responded to market pressures by choosing the chemical agriculture approach. Perkins writes,

> To the farmer, of course, it matters little whether his yields are reduced in quantity or quality. In either case, his revenues are reduced and he is economically worse off. Moreover, no single farmer can decide unilaterally to abandon either quantitative or qualitative criteria in judging production methods. Competition in capital-intensive agriculture encourages each individual grower to strive for maximum returns; those who don't risk losing their businesses. (1982, 268)

The extensive groundwork laid by wartime chemical research and the subsequent profits reaped by chemical companies caused a self-reinforcing cycle of interdependency among the agrichemical industry, researchers, and farmers. There was profit for industry and farmers, and there was research support and recognition for scientists. It would have been difficult for any of those involved to step outside of this relationship and survive economically. This was complicated by the fact that at least with entomological biocontrol "once a predator is released no further costs are required to keep the pest under control" (Morrison 1989, 4). Consequently, for-profit corporations have little incentive to support research and commercialization of entomological biocontrol agents (Ordish 1967, 192).

Chemical domination of agricultural pest control was rocked to a degree in the 1960s by concerns about the environmental damage that chemicals could cause, by evidence of pesticide resistance in pests, and by a trend toward more intensive farming with less crop rotation (Cook and Baker 1983, 41). By the 1970s there was a marked increase in interest in biological control (Greathead 1989, 43, 44). What is perhaps the most well known biological control agent, *Bacillus thuringensis,* was launched commercially in France prior to World War II and has been available in the United States since the late 1950s. According to one recent discussion, however, "it did not become commercially viable until the 1980s when problems of reliability, formulation and application had been overcome, stimulated by the *banning of conventional pesticides* for control of defoliators in North American forests in the early 1970s" (Greathead 1994, 196; emphasis added). Indeed, the case of *Bacillus thuringensis* clearly illustrates the way the context for research on biocontrol agents and their use has been defined by the history of chemical (industry) domination of pest control. The successes of chemical pesticides have determined what is considered to be acceptable levels of pest control. In this context, some analysts suggest that the future of biocontrol will depend on the willingness of farmers to give up the "expectation of perfect activity in favor of adequate control" (Becker and Schwinn 1993, 363). Of course, the constraints of economic competition will not permit farmers to make such a choice individually, but as the recognition of the problems with chemical pest control grows, new standards of adequate control may emerge (see Malakoff 1999a and 1999b).

Whatever the future holds, to understand recent and current work done by biological control researchers, one must see their efforts in light of the inextricably linked historical paths of chemical and biological pest control. We must understand how research and commercial use of chemical pesticides has shaped the design of experiments, the use of research tools, and what constitutes success at an applied level. To address these issues I consider two specific cases. First, drawing on work by Richard Sawyer (1996), I explore biocontrol research undertaken in California, where efforts were made for over half a century to use so-called beneficial insects to suppress harmful insect infestations. Here, cosmetic standards made possible by the use of chemical insecticides set standards of success beyond the reach of most biocontrol agents, and intimate knowledge of DDT made it an important research tool beyond its use on the farm. Finally, I turn my attention to the more compact, but equally rich history of UW85 and the work of the Handelsman lab. Here again, we see the *indirect* effects of chemical industry dominance of pest control on the research and writing of the Handelsman lab.

Success and the Spotless Orange

The early success of entomologists in controlling the outbreak of cottony cushion scale in California citrus groves laid the groundwork for the institutionalization of the first academic and applied research program in biological control in the United States (Sawyer 1996). The California state biological control program owed its solidity and longevity to support from the state's citrus industry. It is perhaps not an exaggeration to claim that the control of cottony cushion scale by the vedalia beetle saved California citrus. The most recent and perhaps most comprehensive chronicler of this history notes, "After the first success of biological control [in California] . . . , growers used their political and economic power to encourage the state, the university, and the scientists themselves to make further efforts. Citrus producers' enthusiasm for biological control led to the specific inclusion of nonchemical methods in the state's agricultural bureaucracy" (Sawyer 1996, xxii; see also Palladino 1996). In this instance the influence of citrus producers on research priorities was quite *direct*. The intervention of citrus industry representatives in the policymaking process is what led the state to support biological control research.

The needs of growers determined the kinds of research undertaken by scientists as part of the state's biological control program. The citrus market was highly competitive, and as the "citrus industry fought for a larger place in the nation's diet," the industry promoted the marketing virtues of pristine cosmetic standards. Insect damaged fruit could not be sold. A 1925 insecticide advertisement directed at growers captures this state of affairs. The ad stated, "No matter how good it tastes looks are what sells" (quoted in Sawyer 1996, 35). In this context, pest control worked not only to maximize production, but also to keep prices high and fruit appealing to the eye (Sawyer 1996, 17, xxii, xxv). Sawyer notes, "Even a few tiny red scales or discolored spots caused by other insects ate into profits, if not into the taste or nutritive value of the fruit itself" (1996, 36). For scientists, this meant finding biological control agents that could permit a spotless citrus fruit to grow to maturity, and this is what California biological control scientists sought in their work: to find insects that would control pests to such an extent that ripe fruit would meet the industry's cosmetic standards. In this case, the guidelines established by the needs of industry *indirectly* shaped the practice of entomological research. The standards for an effective biocontrol agent were not established by explicit demands from industry research patrons, but rather, indirectly, through commercial standards established for saleable produce.

In the years between the late nineteenth century success against cottony cushion scale and the Second World War, there were a number of insect outbreaks in citrus and other commercial crops. California-based researchers sought biological control to address the problems. Operating under the theory that pest outbreaks occurred when pests and their natural enemies were out of balance, and that this occurred when a pest was imported to an environment where there were no natural enemies for that pest, researchers determined that the best place to find and collect natural enemies "was where they had just encountered a previously isolated pest population and had not yet diminished their own food supply" (Sawyer 1996, 14). Typically, this meant scouring the globe to determine where a given pest originated and then finding its natural enemy in that locale and returning it to the place of outbreak.[6] In a number of instances, investigators returned to the United States with predators only to find that while they might substantially reduce insect populations, permitting fruit to ripen and retain its flavor and nutrient value, some insect pests still remained and caused cosmetic damage, thus rendering the fruit unsuitable for the market. In the mid-twentieth century olive industry, even one scale could scar a fruit and make it unsalable. Researchers determined that a parasite from Iran, *Aphytis paramaculicornis*, substantially reduced the scale population as it spread through California olive orchards. But the rigid cosmetic standards meant that though this parasite could control the scale, the level of control was not sufficient for commercial purposes (Sawyer 1996, 178, 179).

Importantly, comparison with the European case makes it clear that standards of adequate pest control and measures of experimental success are not absolute, but quite context dependent.[7] The levels of control researchers were able to achieve through insect-based biological control were insufficient to meet industry needs in the United States. By contrast, similar levels of pest presence were not viewed as a barrier to fruit salability in Europe. Regional variation notwithstanding, cosmetic standards would define what was considered adequate biological control at both experimental and practical levels in the United States.

The developments arising from the Second World War threatened to further weaken the status of biological control research and application in California. The state was "at the forefront of the chemical explosion after World War II" (Sawyer 1996, 158). The government provided lavish support to insecticide research, and growers liberally used the new chemicals. In contrast to chemical use, biological control methods "took too much planning and supervision, and had a low rate of success" (Sawyer 1996, 157). In this context, as pests became resistant to one chemical, rather than questioning what was happening to biological systems, researchers produced new chemicals that were introduced and used. Ironically, however, according to Sawyer, DDT ultimately demonstrated the importance of biological control in citrus. Work on test plots in 1944 and 1945 indicated that DDT would be effective as a means to control for citrus thrips and citricola scales. In 1946, entomologists recommended DDT treatment for these pests; however, application of DDT to 75 percent of the citrus growing area "caused the biggest outbreak of cottony-cushion scale since 1889" (Sawyer 1996, 174). DDT killed the cottony cushion scale's predator, the vedalia beetle, and these bugs had to be reintroduced to control the epidemic. This scenario was repeated on a number of occasions: DDT was used to control one insect, which allowed other species that had not previously been pests to reproduce wildly. Nevertheless, despite the fact that biological control was less likely to cause outbreaks and lead to pests developing resistance to chemicals, insect-based biological control could rarely reach levels of pest suppression that was demanded by industry and already achievable with chemicals. Consequently, by the 1970s, some California entomologists "were calling for relaxation of cosmetic standards [in order] to broaden the applicability of biological control" (Sawyer 1996, 195).

Sawyer notes at several points in his book that researchers in the period before World War II, and indeed until sometime in the 1970s, did not challenge the cosmetic standards established by industry, even though they were aware that less stringent standards would make insect-based biological control a viable pest stabilization technique. It seems to me that Sawyer's concern with this "failing" on the part of researchers misses the point. Biological control research in California did not develop in a vacuum. Rather, it developed in a particular kind of environment in which the rules of the game were beyond the control of any individual researcher or grower. Particular cosmetic standards were set to appeal to consumers. Once established and accepted by consumers, growers who failed to meet these standards were threatened with decreased sales, profit loss, and possibly the loss of their livelihood.

Sawyer further implies that scientists were in the pocket of industry who paid for their investigations. His argument is that industry influence was *immediate* and *direct,* by serving as patrons for the university and for state scientists, industry set research agendas and research standards. To a degree, this is true. I would emphasize, however, *indirect* and *structural* factors. As the rules defining industrial practice exist outside the control of any individual grower, they are also beyond the capacity of individual scientists to alter them. If scientists wished to undertake biological control studies that would be meaningful to industry, even if their studies were not funded by industry—to do work that could ultimately be applied in citrus groves that would then produce saleable fruit—their tests of biological agents had to meet control standards acceptable to industry. In this way, the very notion of successful scientific investigation was shaped *indirectly* by the logic and demands of an industry.

This is not to suggest that the rules of the game—in this case, the cosmetic standards for fruit—are in any way ahistorical or trans-historical. To the contrary, they were clearly established at a particular point in time but once they were firmly entrenched, these guidelines fundamentally shaped the practices of growers and scientists alike. To change them would have demanded collective action by consumers, growers, or, less likely, scientists under particular historical conditions (that is, some sort of crisis).

In addition to the role played by the citrus industry in defining what counted as successful biological control in entomological research in California, it is worth noting another indirect and perhaps unanticipated effect that the agri-chemical industry had on biological control research in California in the period after World War II. If cosmetic standards that were unattainable by most biological control agents constituted a hindrance to successful research by agricultural scientists, DDT had an interesting and beneficial effect on entomological biocontrol after the War. There was, of course, the fact that DDT prompted insect outbreaks, potentially pointing to the limitations of the chemical paradigm and the possible utility of biological control of insects. But beyond this, DDT could also be used as a tool in biological control investigations. According to Sawyer, one entomology researcher

> turned DDT into a tool for demonstrating the value of natural enemies of citrus pests. It had always been difficult to evaluate natural enemies. They could fly from test plots to control plots and spoil experimental results. [However, s]ince DDT killed many predators and parasites more readily than it killed pests, DeBach could use it to exclude natural enemies. Treated plots became controls in what [this scientist] called the "insecticidal check method." Lower pest populations in untreated populations indicated the effect of natural enemies. (1996, 174)

Such natural enemies could then be artificially introduced to control pests.

Thus, at least at two levels, the braided history of industry—here the citrus and agri-chemical industries—and biocontrol research shaped the very fabric of research. At one level, standards of research success were indirectly established by measures of commercial success. At another level, a chemical—DDT—that finally came to be perceived as a rather ambiguous good in the realm of the environment and the market, proved to be a useful research tool for understanding the efficacy of specific biological control agents.

B. cereus: UW85 and Chemical Fungicides

Where do we place UW85 in the intertwined history of chemical and biological control of plant pests? Handelsman lab research on UW85 shows that this microorganism—a strain of *Bacillus cereus* (or *B. cereus)*—is effective in controlling certain soil borne diseases caused by a microorganism called *Phytophthora.* Work on this problem must be understood not in the context of insecticide development and use, as was the case with the research on the pests in California citrus groves, but rather in the context of fungicide development and use.

Fungicides were sold at a rate of some $595 million in 1995. This is up 14 percent from 1994. Although fungicides have diverse uses in many areas, including the lumber, paint, plastics, and pharmaceuticals industries, the agricultural industry is thought to be the largest consumer of these chemicals. In agriculture, treatment for fungal diseases is heaviest in fruits and vegetables, followed by peanuts (*Industry Surveys* 1997). Fungicide use is especially concentrated in Appalachia, the Delta region, and the southeastern United States. According to one source, "These regions generally have greater disease problems than other regions because of climates that are conducive to pathogen growth" (Osteen and Szmedra 1989, 26).

Other regions and crops are also susceptible to soil borne disease, especially where drainage is inadequate. In Wisconsin—where the Handelsman lab is located—and the north central United States more generally—alfalfa is an important crop. It is a feed source for Wisconsin's $3.8 billion dairy and beef industries, and adds $72 million to the state's hay crop. According to one Handelsman and Goodman grant proposal, alfalfa "underpins" some 50,000 jobs in areas as wide ranging as plant breeding and dairy farming (Goodman et al. n.d., 1). Alfalfa is threatened in certain regions by *Phytophthora* root rot, one of the most destructive diseases affecting that crop.

According to one source,

> the technical difficulties associated with controlling soil-borne pathogens do not encourage the search for new chemical solutions. Soil pathogens often have special survival structures or strategies . . . or they persist in crop residues which allow them to remain alive and infectious for long periods of time. Their distribution throughout the soil makes it impossible, for ecological as well as economic reasons, to control them directly with a fungicide. (Becker and Schwinn 1993, 357)

Indeed, between the 1960s and 1983, cultural and physical practices, as well as the use of resistant varieties were the primary means farmers used to avoid disease devastation of alfalfa. In the 1950s and early 1960s, Thiram and captan were used in alfalfa fields. Use of Thiram declined because it killed microorganisms important for nitrogen fixation and thus plant health. The use of

Thiram led to little yield increase, and it fell out of favor (Craig Grau, correspondence with the author, December 15, 1997).

In the last twenty years a new fungicide effective against root rot has been developed. Metalaxyl—an acylalanine fungicide—was "discovered" by researchers at the agri-chemical company Ciba-Geigy in 1977 and approved for widespread use as a fungicide in 1983. Metalaxyl is "target specific," meaning that it is effective against specific fungi that "incite" specific diseases. The anti-fungal agent is active against *Phytophthora, Pythium,* and *Peronospora* (Carley 1994, C22) and, consequently, metalaxyl is not only used to treat alfalfa, but also marketed for soybeans, apples, broccoli, cabbage, cauliflower, cucumbers, onions, spinach, and tomatoes (Marshal 1984, 10).

Thus, it is in the context of commercially produced fungicides, and metalaxyl in particular, that research on UW85 should be understood. In the discussion that follows, I focus on four ways that the practices of Handelsman lab scientists are affected by the history of pesticide use in agriculture and the development of metalaxyl. First, I consider how this history has significantly affected the ways in which academic papers and grants produced by the lab and its collaborators are framed and justified. Next, I explore how the existence of commercially available fungicides for the same diseases that UW85 controls influences the way field experiments are organized. Third, I discuss how research results on UW85 are reported in the context of the existence of metalaxyl. Finally, I describe how, like DDT in entomological biological control research, metalaxyl is used as a research tool by the Handelsman lab.

Professor Handelsman and her colleagues published their first major paper on UW85 in 1990, in the journal *Applied and Environmental Microbiology*. In the introduction to this article, Handelsman and her coauthors point out that *Phytophthora megasperma* f. sp. *medicaginis* is a cause of seedling and root disease in alfalfa, which can substantially reduce yield. They argue that resistant varieties of alfalfa do not offer sufficient protection against these diseases, and they suggest that while the fungicide metalaxyl offers additional protection, even its efficacy is limited: "a low frequency of isolates of *P. megasperma* are resistant to this fungicide, suggesting that its utility may ultimately be limited by selection for a resistant population of *P. megasperma* f. sp. *medicaginis*"

(Handelsman et al. 1990, 713). As a consequence, according to the authors, "Multiple control measures may be required to obtain optimum yields in the presence of the pathogen complex" (Handelsman et al. 1990, 713). Among these would be the use of a new biocontrol agent. Thus, from its very beginnings, research on UW85 was justified in relationship to the use of chemical disease control: it is the inadequacy of metalaxyl that provides part of the justification for the search for a biocontrol agent.

In a more recent paper describing genetic resistance to antibiotics, the justification for UW85 over a chemical alternative is more complicated. According to Handelsman and her coworkers, "Biocontrol has been hailed as a control measure that will be *more robust* than synthetic chemicals" (Stohl et al. 1996, 475; emphasis added). However, the authors continue, even though resistance to antibiotics is likely to occur in pest targets of biocontrol agents, "antibiotic resistance is likely to be less of a problem in biocontrol than in pest pathogen control strategies based on synthetic chemicals, since biocontrol agents often produce multiple antimicrobial compounds, making it less probable that the pathogen will develop resistance to the entire suite of inhibitory molecules involved in biocontrol" (Stohl et al. 1996, 475).

In a 1995 grant application, Handelsman and her colleagues justify their work on UW85 and two other *B. cereus* strains by pointing out, first, that alfalfa underpins the dairy, beef, and racehorse industries in Wisconsin. The proposal suggests that alfalfa is vital to the Wisconsin economy, but that despite pest resistant varieties and availability of fungicides "additional approaches to root rot disease control are needed" (Handelsman et al. 1995, 4). Among the reasons new approaches are necessary: "the limited effectiveness of present genetic resistance to the total disease complex [from which alfalfa can suffer], the existence and potential for selection of pathogen strains that infect resistant genotypes, the general failure of available resistances to protect very young plants, chronic disease of fibrous roots and/or nodules, and the environmental costs, variability, and re-registration issues associated with disease control by fungicides" (Handelsman et al. 1995, 4).

This final issue the grant writers raise concerning environmental costs and possible problems with government registration of certain fungicides points to a different aspect of the braided history of chemical and biological control than the lab's first paper, in which it was asserted that the fungicide simply is not good enough. In this case, the writers draw attention to the changing climate in which pest-control research is undertaken. Today, in light of research and political efforts that have pointed to the possible environmental hazards posed by chemical pesticides, alternatives that do not pose these threats are viewed as valuable. The implication here is that UW85 and its *Bacillus cereus* relations may be more environmentally friendly than chemical alternatives.

In a 1996 paper by Handelsman and members of the Goodman lab, their justification is less direct. Biological control is simply referred to as an alternative to chemical control. In the paper's abstract, the authors contend that "Systematic acquired resistance (SAR) and microbial biocontrol each hold promise as alternatives to pesticides for control of plant diseases" (Chen et al. 1996, 73). This justification is developed further in the body of the paper:

> In agriculture, seedling damping-off diseases, many of which are caused by Oomycetes, are a significant factor and the justification for substantial use of pesticides. Laboratory and greenhouse studies have shown that SAR or the expression of SAR-related transgenes may offer promising alternatives or adjuncts to pesticide use for control of these pathogens. . . . Likewise, microbial biocontrol with UW85 shows promise in both laboratory . . . and field . . . studies, for control of seedling damping-off and rot root diseases caused by Oomycetes. In both cases, however, the results thus far reported suggest that, *as is sometimes the case with pesticides,* control is variable (Chen et al. 1996, 78; emphasis added).

It is interesting to note here that although the efficacy of biocontrol and SAR is qualified, the authors point out that chemical-based disease control can also exhibit the kind of variability that could limit the usefulness of these technologies.

One might conclude that this discussion of the Handelsman lab's papers and grant applications merely illustrates the recognition by lab members of the importance of rhetoric in science (see Myers 1990; Gieryn 1999; Kleinman and Solovey 1995). However, the appropriateness of the language and imagery used in the writing produced by the lab is defined by a particular context. In promoting its work, the Handelsman lab frames its research in terms that have currency among colleagues. In a field that exists in the shadow of chemical pest control, it is almost inevitably the case that comparisons between biological and chemical control are made. Furthermore, this framing is integrally related to the work done by the laboratory at the bench, in the field, and in the greenhouse, where chemical pest control affects experimental design, measures of success, and research tools.

In this context, I turn to the laboratory's experimental work. Since 1986, the Handelsman lab and its industrial collaborators have taken UW85 to the field to test its disease fighting capabilities. In the summer of 1996, the lab had two soybean trials, two alfalfa trials, and an ongoing alfalfa trial from the prior year at three sites; all trials were in Wisconsin. The lab's industrial collaborators have sponsored or themselves undertaken a number of other trials around the country on such crops as wheat, rice, peanuts, and Brassica cold crops. According to Handelsman, there were years when there were as many as twenty-five field trials going on around the country (interview July 25, 1996). Typically, in these field tests the efficacy of UW85 and related *B. cereus* strains is measured against metalaxyl or some other commercially produced fungicide. In a sense, the biocontrol agent's success as a disease suppressant is measured against a chemical yardstick.

In the lab's earliest small-scale field trial, alfalfa seeds were either coated with methylcellulose (equivalent to no treatment), methylcellulose and UW85, or metalaxyl. Here, seedling emergence was measured, and UW85 performance was comparable to the metalaxyl treated seed (Handelsman et al. 1990, 716).

In field tests carried out in Marshfield, Wisconsin, in 1993 and 1994, the lab grew alfalfa and plots were divided between those in which no disease suppressant treatment was provided, those treated with metalaxyl, and those treated with UW85 and two other *Bacillus cereus* strains. Performance was measured based upon seedling emergence and total forage yield. UW85 did not perform as well as metalaxyl on either count, but the other two strains of *B. cereus* were comparable or superior to metalaxyl (Handelsman et al. 1995).

In a 1991 field test reported in the journal *Biological and Cultural Tests for Control of Plant Diseases* the organization of the experiment was similar. This study was performed by P. M. Phipps of the Tidewater Agricultural Experiment Station and the Polytechnic Institute and State University of Virginia in Suffolk. The aim was to ascertain the efficacy of biological agents in controlling sclerotinia blight of peanuts (Phipps 1992, 60). Biocontrol agents were compared with the registered fungicide treatment, Rovral+Nu-Film-17 and an experimental fungicide, ASC-66825, or fluazinam. The plot treated with UW85 produced a yield of 3,504 pounds. This compared favorably with the registered fungicide treatment, which produced a yield of 3,258 pounds; the experimental fungicide, however, out performed UW85, producing a yield of 4,577 pounds.

How research results are reported in academic journals reflects the structure of field trials. According to a peer reviewed scientific article published by Handelsman lab workers and their industrial collaborators, "UW85 treatments provided a yield benefit each year that was *of similar magnitude* to the benefit provided by Ridomil," which is one of the trade names under which metalaxyl is sold (Osburn et al. 1995, 555; emphasis added). In further comparing UW85 and metalaxyl, the report makes an additional important observation: "Historically, yield enhancement by bacteria [such as UW85] that suppress disease and promote plant growth have been plagued by *inconsistent performance* in the field. In the five-year trial described here, we observed *no greater variability in performance* of UW85 than in that of a commercial fungicide" (Osburn 1995, 555; emphasis added).

This type of comparative testing and reporting of results is not uncommon in biocontrol research, if work appearing in the journal *Plant Disease* during the 1990s is any indication. In one study, a technique called "Bio-priming" is compared to metalaxyl in controlling damping off in sweet corn. The investigators report that Bio-priming with *P. flourescens* AB254 "provided protection against damping-off as good as or better than seed treatment with metalaxyl when seeds were planted in cold soil" (Callan et al. 1990, 368, 370, 371). Similarly, a paper on Chickpea Wilt reported that "In one of the [field] trials, the yield from *P. flourscens* treatment . . . was much higher than the yield after carbendazim application itself, while in the other it was on par with the fungicide application" (Vidhyasekaran 1995, 784). In one final example of research on seed rot of chickpeas, researchers reported: "Moderate protection against Pythium seed rot in the field was obtained with the three biological seed treatments. *P. flourescens* strain Q29Z-80 provided the best yield in the field trial and was statistically equivalent to captan, with respect to emergence and plant height, and with the steam soil control with respect to plant fresh weight" (Trapero-Casas et al 1990, 568).

Metalaxyl is not only a yardstick used to measure the efficacy of UW85, but similar to the use of DDT in citrus-insect research, is a useful research tool for the lab. Handelsman describes the role metalaxyl can play in research in the following way:

> When plants don't emerge or die in the field, it is hard to determine why. Was it disease or something else? If it was disease, which pathogen caused it? Using a highly specific chemical is a powerful way to discriminate among possible causes. If emergence is significantly higher in the metalaxyl treatment than in the control, then the Oomycetes pathogens are likely to be involved since metalaxyl is pretty specific for the Oomycetes. (e-mail correspondence, May 4, 1997)

Again, use of a commercially produced chemical as a research tool is not restricted to the work of the Handelsman lab. In the sweet corn study discussed above, for example, researchers used existing chemical treatments to determine that "neither insects nor fungi other than *Pythium* spp. were involved in problems of seedling emergence" (Callan et al. 1990, 369).

The extensive history of chemical pesticide research is, in this instance, a benefit to the work undertaken by the Handelsman lab. Ultimately, in contrast to cases in which work on UW85 is justified by its commercial potential over chemical fungicides, this tool would be useful to the lab even if metalaxyl were taken off the market.

The historical legacy that places metalaxyl where it is in the work of the Handelsman lab is not lost on Handelsman herself. She notes, "Many billions of dollars have been poured into chemical research over the last 50 years. . . . Therefore, our knowledge base with chemicals is much greater [than that of biological control]. I wonder how we would conduct our experiments differently and how our knowledge base would look if 10 billion dollars had been pumped into biological control" (e-mail correspondence, July 9, 1995).

It is interesting to note that Handelsman is alert to further commercial justifications for the development of UW85, and that these justifications are linked to chemical alternatives. First, as reported in a soybean producer's trade paper, "Only a small amount of the organism [UW85] gives fairly sustained disease suppression because it grows and spreads on soybean roots. For that and other reasons, Handelsman predicts that the biological control agent *will cost less than commercial fungicides*" (Behling 1993, 34). Among the other reasons Handelsman suggests UW85 will have an important market advantage over existing chemicals is that seed treated with chemicals cannot be fed to livestock or humans. According to Handelsman, this a problem because often farmers have unused seed at the end of the season. If UW85 were approved for animal and/or human consumption, it would have an advantage over chemical treatments because unplanted seed could be sold as feed. According to one of the lab's industrial collaborators, farmers are willing to take the yield loss associated with the diseases rather than end up with unusable chemically coated seeds (interview, July 25, 1996).

Conclusion

A good deal of research in the history of science and technology looks at the *direct* role that patronage has in shaping research questions and the character of scientific fields. David Noble (1984), for example, explored the role of military patronage in promoting the development of numerical control machine tool technology over record playback equipment. Stuart Leslie (1993) focused on the role of military support for university science in shaping the field of materials science and promoting specific research methodologies. Robert Kohler (1990) traced the influence of Rockefeller Foundation funding on the contours of the field of molecular biology.

As I discussed in chapter 2, over the past two decades the emphasis on the impact of direct financial support from industry to university scientists has also been central to discussion about the costs and benefits of university-industry relations (UIRs) in the biological sciences. Nelkin and Nelson express concern that corporate funding will alter the research priorities of university scientists (1987, 71). This is a fear also expressed in a 1983 American Association of University Professors' report (1983, 21a). Others worry that corporate funding of university science can undermine the traditionally open character of scientific communication (see Kenney 1985; Shenk 1999).

All of this research draws our attention to the *direct* influence of patronage on the character of academic science; this is an issue of particular importance as we begin a new century and a period in which the federal government cannot meet the demands to support increasingly high cost research. But concern with these matters should not lead us to underestimate the *indirect* effects powerful institutions exert in shaping the character of scientific practice, and, indeed, a complete understanding of university science today demands that we examine such factors.

I have emphasized these issues in this chapter. Research on chemical control of agricultural pests might be said to have begun in the late nineteenth century, but it gained substantial momentum with the two World Wars. Even in this early period, we can say that much of the influence of the chemical industry, as well as of the federal government on biological control research was *indirect.* Some have suggested that the relative neglect of biocontrol studies in these years may have stunted the field's development. But the pervasive and *indirect* effects of chemical industry dominance of agricultural pest-control research are also clear in the ways scholarly writing in the field is framed, the way experiments are organized, the measures of success that are used and the tools that are available. In this chapter, I illustrated these indirect effects by first outlining the history of pest-control research and looking at the relationship between research on chemical and biological pest-control agents. I then explored biocontrol in the citrus industry. Here, drawing on the work of Richard Sawyer (1996), I showed how cosmetic standards affected definitions of experimental success and how DDT—a product of WWII research—was used as a research tool in citrus biological control work. Finally, I turned my attention to the work of the Handelsman lab on UW85, showing how the fungicide metalaxyl is used in field experiments as a standard of comparison for the lab's biocontrol agent, and how metalaxyl is used by lab workers as a research tool. In both the citrus and UW85 cases, industry affected scientific practice not directly through research funding, but indirectly through institutionalized standards and tools developed through earlier industry-supported research.

Regarding the casting of power I presented in the previous chapter, we can see explicit cases in which the deep pockets of industry allowed it to directly shape the research terrain for agricultural pest protection. Industry funded some and not other research, and in the citrus case, it established cosmetic standards for the commercial sale of fruit. In so doing, to use the language of early power theorists, A's action (industry's) prompted B (scientists) to behave in ways that B might not otherwise behaved. This explicit research support and direct action by industry provided the basis for a different dimension of power. This established situation influenced the opportunities and constraints faced by researchers. Once these standards and tools were established, no direct intervention or explicit exercise of power by industry needed to occur for the influence to be felt. To turn again to Steven Lukes, in this instance, the bias of the agricultural research system "is not sustained simply by a series of individually chosen acts, but also, most importantly by the socially structured and culturally patterned behaviour of groups and practices of institutions" (1974, 21, 22). Whereas in the early history of agricultural pest-control research power was initially exercised primarily through the direct actions of organizations, ultimately, as in the cases I have discussed, to a significant degree power became an attribute of structures, and direct action by powerful actors was not necessary for power to have effects. Direct and explicit exercise of institutionalized power provided the structural basis that then produced pervasive effects.

The extent of these indirect influences on academic science is difficult to measure. One can imagine, however, that similar forces are at work in areas like transportation research and energy studies, fields in which any alternative inventions must compete with existing products. Beyond cases that roughly parallel the development of biocontrol research, the direct effects of patron support for academic science continue to have indirect influence in shaping the character of the research terrain long after direct support ends.

4

(Un)Intended Consequences

Commercially Produced Research Materials and the Transformation of University Biology

The techniques and approaches of molecular biology pervade life science research in virtually all fields (see Fujimura 1996). Although analysis and experimentation in the Handelsman lab takes place at a range of levels, from farm fields and soil ecology to bacterial cultures and biochemical interactions, the techniques of molecular biology are central to efforts of workers in the lab. Even when these molecular biology tools and techniques are not used, the research the lab undertakes still qualifies broadly as "big science" biology. Reagents, equipment, and testing services, not to mention staff and administrative costs, are all expensive. As I have noted, in the mid-1990s the Handelsman lab required approximately a quarter of a million dollars annually to undertake its work.

The standardization of research tools in biology—especially molecular biology—has increased the pace of science dramatically. Scientists make reputations on the development of faster techniques (Fujimura 1996, 77). Indeed, papers from the Handelsman lab point to the efficiency of new assays and methods developed in the lab. For example, in one paper lab members describe a rapid method for the purification of zwittermicin A (Silo-Suh, Stabb, Raffel, and Handelsman 1998). According to a 1997 news report, the aim of one research materials company is "to speed the research process by sparing labs the time-burning task of creating their own cDNA libraries or labeling reagents for automated DNA synthesis or Northern blots used for cancer research" (Transpacific Media, Inc. 1997, 56). The catalogs of biological research materials companies also promote new products with claims of significant increases in *efficiency*. A 1997 Epicentre Technologies product catalogue refers to their new PCR Enhancement Technology that "dramatically improves the yields and template-to-template consistency of PCR." Invitrogen's 1997 catalog suggests product consumers try their new transfection kit and "find out how you can achieve higher mammalian cell transfection efficiencies" (1997, vii). Similarly, advertisements that are splashed through *Science*, the weekly journal of the American Association for the Advancement of Science, promote technologies that provide more product more quickly. An advertisement for Invitrogen's InsectSelect™ System says the system "allows continuous expression of recombinant proteins, saving . . . time, effort, and expense." The system, the advertisement continues, includes "a c-terminal fusion tag for efficient purification and detection of proteins" (*Science*, April 14, 2000, p. 219).

The fact that research tools of these kinds have become indispensable to the practice of contemporary science has not been lost on science studies scholars. As I have discussed, Latour and Woolgar make the analysis of inscription devices central to their study of laboratory life. They contend that the phenomena studied by the lab workers whom they observed could not exist without inscription devices. Indeed, according to Latour and Woolgar, "the phenomena *are thoroughly constituted by* the material setting of the laboratory" (1986, [1979], 64). In addition, they explore the way research tools developed in a laboratory come to be unexamined black boxes that are often available through commercial firms.

More recently, Clarke and Fujimura made research tools a central issue in science studies with their edited volume, *The Right Tools for the Job* (1992).[1] Emphasizing her view that social worlds cooperate, Fujimura herself suggests that technical standardization provided "a means through which techniques could be transported between labs" (1996, 6, 7).[2] More generally, Fujimura states that "standardized packages"—a blend of tools and theory that have become widely accepted—provide the basis for an environment in which cooperation among diverse "social worlds" is possible (Fujimura 1996).[3]

As I have stressed, these and other analyses add a great deal to our understanding of scientific practice. Certainly the idea of an inscription device helps clarify how the material scientists use is lost in the process that establishes scientific facts. It is also clear that modern science could not function without the standardization of tools and the resulting cooperation that Fujimura describes. But such analyses lead us to overlook some important implications of the pivotal role of research tools in the life sciences. Most fundamentally, relationships between university laboratories and research tools suppliers are often not thoroughly examined.[4] That research tools are often purchased from companies is mentioned in many science studies texts (see Knorr Cetina 1981; Latour and Woolgar 1986 [1979], 61), but often with little exploration of the implications of this relationship. In particular, I believe, we would benefit from a discussion of the power and resource asymmetries that characterize such relationships as well as consideration of the practical and pedagogical pitfalls of pervasive standardization and use of black boxes.[5] In this chapter, I consider the views of some members of the Handelsman lab regarding the benefits of the new and often commercially produced research tools which they now use. However, the central focus of the chapter is on the *unintended* (and often negative) consequences of the reliance of academic biology laboratories (specifically the Handelsman lab) on standardized, primarily commercially produced, research tools.[6] I consider how these unintended consequences are entangled with questions of power and resource dependency.[7] After providing some general background, I discuss three different cases. First, I consider how a relationship of resource dependence complicated by conflicting understandings of reputation shaped the interaction between the Handelsman lab and a company with which they contracted for a testing service. Next, I consider how the black-box character of a commercially produced research tool used in a project undertaken jointly by the Handelsman lab and that of Robert Goodman influenced where scientists looked for contamination. Finally, I explore the implication of the use of "kits" in life science research for the education of scientists-in-training.

The Life Sciences Research Tool Market

Molecular biology came of age in the 1970s when it was demonstrated that genetic manipulation through the techniques of recombinant DNA (rDNA) was possible (see Wright 1986; Fujimura 1996; Judson 1992). But historian Susan Wright points out, "Molecular biology in 1970 represented the culmination of a research tradition that originated in the 1950s in the merging of structural, biochemical, and informational approaches to the central problems of classical genetics" (Wright 1986, 305). The model of DNA proposed in 1953 by James Watson and Francis Crick provided the foundation for this tradition. Their framework suggested that genes are made of DNA, which encodes information that in turn determines the processes of genetic replication and protein synthesis.

Industry was involved in the development of research tools as early as the nineteenth century, but these were not standardized tools. Instead, firms developed "research tools tailored to specific needs, local markets" (Gaudilliere and Lowy 1998a, 6). In the twentieth century, the individual instrument-maker was replaced by the industrialist "either as manufacturer of mass-produced devices (instruments, reagents, living organisms) or as a supplier of individualized products geared to the research market" (Gaudilliere and Lowy 1998a, 6). In the biological sciences, standardized tools were used even before the dominance of rDNA techniques and technology.[8] As early as 1929, the Jackson Laboratory in Bar Harbor, Maine, developed animal research colonies for the study of mammalian genetics (Fujimura 1996, 26). Through inbreeding, the lab aimed to produce a group of animals with the highest possible genetic similarity. Although the Jackson Lab and others like it sold research animals to investigators not associated with the labs, "Commercial supplies of inbred animal strains (and their professional animal breeders) were not available until after World War II" (Fujimura 1996, 28).

By the 1950s and 1960s, tissue culture techniques for use with multicellular systems had been standardized and had become routine tools for a large number of researchers (Fujimura 1996, 39). And it was in the 1960s that a crucial tool in molecular biology was discovered: restriction enzymes were believed to play an important function in protecting bacteria from foreign DNA

(Wright 1986, 305; Judson 1992, 62, 63). These enzymes made it possible to cut DNA in exact locations; it was quickly learned that they could be used to cut and join pieces of DNA from different sources that would not normally be combined in nature.

According to Fujimura, "by the 1980s, despite claims that recombinant DNA technologies were states-of-the-art methods, these technologies had been transformed into relatively routine and standard protocols" (1996, 71). The steps in such techniques can be preformatted, and instruments often permit a set of tasks to be undertaken automatically. Standardized technologies reduce the amount of tacit knowledge and discretionary decision-making required to solve many routine problems (Fujimura 1987, 266). Indeed, by the 1980s, the routine character of these techniques and tools was sufficiently complete that the "gifted hands" of a recombinant "artist" were often unnecessary. When I joined the Handelsman lab in 1995 with no experience in molecular biology laboratory technique, I was taught to perform a standardized assay. Undergraduate technicians with limited knowledge of biology could also be taught to undertake experimental work. In fact, by the 1980s, a good deal of research in molecular biology progressed without the custom-made tools common in some fields.[9]

In a random sample of one month of articles appearing in the *Journal of Molecular Biology* in 1972 and one month appearing in 1992, I found a dramatic increase in the number of materials that the articles indicated had came from companies, a near doubling from 42 in 1972 to 79 in 1992. The articles indicated that a similar number of materials were made by the lab conducting the research, but the number of materials supplied by other non-commercial labs dropped from 40 in 1972 to only 12 in 1992. As indicated in these articles, the number of machines used in this research nearly doubled over the period, and the number of pieces of equipment that were clearly commercially produced increased three-fold. The categories of lab-produced and commercially produced research materials and tools are sometimes difficult to distinguish, and to a degree these differences in counts may be due to changes in reporting norms. Nevertheless, the vast change in number, plus other research that suggests an increase in standardization in molecular biology, as well as the impressions of biologists with whom I have spoken, all suggest that my data represent a general trend.

In this environment, a vast market for research tools for biological study has developed. According to one source, in the mid-1990s, the worldwide market for specialty research biochemicals was valued at about $1.5 billion (Shon 1994). In 1994, the lab chemicals market in the United States was estimated to be worth from $750 to $800 million (Shon 1994). Universities account for about 25 percent of expenditures of these materials. These figures do not include the cost of equipment. Annual expenditures for equipment used in the life sciences in academic institutions jumped from over $200 million in 1983 to over $400 million a decade later (National Science Board 1996, 187).

The largest producer of research materials for the biological sciences in the late-1990s was Maryland-based LDI, with annual sales of $320 million. Promega has sales of around $60 million making it the second largest company in the field. Promega was followed by La Jolla–based Strategene, with sales of approximately $45 million (Transpacific Media, Inc. 1997, 56). Scientists increasingly turn to these companies for their tools. One analyst noted, "Instead of purifying or constructing their own materials, scientists have switched to mail-ordering of products such as restriction enzymes, modifying enzymes, and vectors required for DNA cloning" (Fujimura 1996, 91). If the image of a tool suggests a stainless steel implement of some sort or even a simple test tube, an article on Clontech, the fourth largest molecular biology supply house in the United States, belies this image: "what Clontech typically sells are eyedropper-sized 2cc vials of exotic solutions with accompanying instructions at $500 to $600 a pop" (Transpacific Media, Inc 1997, 56).

Culture Clash: Analyzing Fatty Acid Analysis

The era of big biology has not brought equipment to rival the size and cost of the great atom smashers used in experimental physics (see Traweek 1988; Knorr Cetina 1999; Crease 1999), but, as I have described, the cost of doing biology today is high; university labs cannot always afford the technology they need to participate effectively in their research area.[10] Sometimes one alternative to purchasing a piece of expensive equipment is to send biological samples to a company that owns the necessary equipment, and will do analysis of these samples on a price per unit basis. Fatty acid analysis of bacterial isolates can be done this way. The Handelsman laboratory's work on UW85 and zwittermicin A led lab researchers to seek such a service in one instance. A conflict that arose between the lab and the contracting company points to a clash of cultures that can emerge as the boundary between the university and the world of commerce blurs. In this instance, a resource dependence relationship was complicated by divergent interests and conflicting notions of reputation.

In the early 1960s, Abel and his colleagues were the first to show that so-called fatty acid analysis by gas-liquid chromatography could be successfully used to identify bacteria (Smith and Siegel 1996, 179). However, while these researchers identified different cellular fatty acid patterns within various bacteria, they did not attempt to identify specific fatty acid profiles. It became clear, however, that bacteria grown under controlled conditions exhibit fatty acid profiles that can be used for identification purposes. Qualitative and quantitative differences in fatty acid profiles permit the identification of distinct genera, species and sub-species of bacteria. According to the website of one company that performs fatty acid analysis, "Whole cell bacteria extracts contain straight and branched fatty acids with 9 to 20 carbon lengths (saturated and unsaturated). The presence of over 330 different compounds along with their quantitative composition yields a wealth of taxonomic information" (www.baronconsulting.com/fsl.htm). Fatty acids are named according to the number of carbon atoms they contain, the types of "functional groups" and the location of "double bonds" (Smith and Seigal 1996, 180).

This analytic process has been fully automated. The system combines chromatographic analysis of bacteria with a computer technology, which includes databases and software that permits comparison of the bacteria under study with a library of data on other bacteria. The producers of one automated system described it:

> The automatic sampler allows a maximum of 100 samples to be run without intervention. After injection of the sample, the temperature increases from 170c to 270c (increasing about 5c per minute) and is driven by the computer program. The resulting peaks are identified based on their comparison to the retention times of a mixture of

known FAs [fatty acids] . . . ; the calibrator data are used to calculate an equivalent chain [carbon] length . . . for the molecule. The peak area amount of each CFA [cellular fatty acid] detected is calculated as a percentage of the total area of CFAs. . . . [A] multivariate statistical method . . . is used as the basis for interpreting data and matching an unknown sample with database entries. (Smith and Siegel 1996, 182)

A former Handelsman graduate student explains the process a bit less formally:

Every different kind of bacteria—all bacteria—have fatty acids as components of their cells—part of how the cell makes its lipids and what not. . . . And you can extract those fatty acids and put them through a gas chromatograph machine, and they'll—the different fatty acids—will come out at different times and make different peaks. So, what people have learned is that different species and even different strains of the same species have different profiles—fatty acid profiles. Much like a finger print kind of profile. And so it is a very powerful technique for asking questions like "well—I've got all these strains. Are they really *Bacillus cereus?* They look like *Bacillus cereus* on plates, but are they really?"

As it turns out, this is an open question. The former Handelsman scientist continues,

The definition of what a species is is rather murky. For instance, *Bacillus thuringiensis* and *Bacillus cereus* are considered two separate species, but they really ought to be the same species by all kinds of tests. They look just the same, and in fact, fatty acid analysis is one of those tests where they . . . [appear] the same. So . . . you can't say at a species level it will tell you definitively, but it will definitely put it in that group. You would know it's not *E. coli*. You'd know it's not even *Bacillus subtilis*. So—it's definitely a much firmer test than say . . . plating on selective media and then doing [plating of] the blood [on agar media]. . . . Actually, that's really pretty effective, but this is definitely a step more sure than that.

Cellular fatty acid analysis has a wide range of uses. The pharmaceutical and food industries can use the method to determine the source of contaminants. Indeed, fatty acid analysis has been used to identify bacteria isolated from food packaging and board machines (Vasanen and Salkinoja-Salonen 1989), and it has been used in clinical settings (Saxegaard, Baess, and Jantzen 1988). In the agricultural area, researchers investigated the failure of a commercial spore larvicide to control the Japanese Beetle. The product analyzed was reported to utilize a particular bacteria; cellular fatty acid analysis revealed, however, that the particular bacteria was absent from the product. Most batches contained different bacterial types than those promoted (Smith and Siegel 1996, 197).

The diverse range of possible uses of the technology make CFA very valuable. But the technology is expensive, and researchers in universities and industry frequently turn to a limited number of companies that undertake fatty acid analysis on a contract basis.[11] In the mid-1990s, these firms would do such testing for between $45 and $60 per isolate, depending on the number of isolates analyzed. In 1997, the Ohio Cooperative Extension Service offered farmers and home gardeners an even more attractive deal: $15 per isolate (UPI 1997).

A former Handelsman lab graduate student found a use for fatty acid analysis in one of his projects. The student had a collection of what he believed were *Bacillus cereus* strains. He had grown them on plates that had "selective media," which theoretically would facilitate the growth of *B. cereus* and not a wide range of other bacteria. But the lab wanted to make certain that "we were really getting *Bacillus cereus*." They wanted to assess these strains using fatty acid analysis, but they lacked the technology in the lab. They "looked around trying to find people that had the equipment and [realized] it wasn't going to be easy to find somebody on campus that . . . could do this." As a consequence, the lab sent the samples to a company to have the work done. The analysis came back and indicated that the samples were, indeed, *B. cereus,* but as the Handelsman scientist studied the fatty acid profiles he discovered that they provided additional information: "I realized that there were a couple fatty acids in the profile which were indicative of whether a strain's producing zwittermicin or not. So all the zwittermicin A producers had a profile that was slightly different than the isolates that weren't producing. And this was on a very limited set of strains." The researcher was on to something: an *efficient* means of finding zwittermicin A producers. He already knew that sensitivity to the so-called P7 phage (the equivalent of a virus in the bacterial community) as well as the ability of certain *Bacillus cereus* strains to inhibit *Erwinia herbicola* were "highly predictive of zmA production" (lab correspondence to company). But neither of these assays or tests identified all zwittermicin producers. Given that preliminary lab research indicated that this novel antibiotic played a role in biocontrol, fatty acid analysis could facilitate the lab's search for new biocontrol agents, perhaps agents better suited to the region in which they were initially found (see chapter 6).

With these results, lab members wanted to do more fatty acid analyses. The graduate student was "kind of hoping we could get one of the machines to do it," but the cost was prohibitive. He estimated the cost to be approximately ten to twenty thousand dollars. In fact, according to a trade journal article published at around that time, purchasing such technology would have cost the lab closer to $50,000 (*Food and Chemical News* 1996, 38). This represents between a fifth and a sixth of the lab's annual budget.

At about this time, one member of the Handelsman lab attended a seminar sponsored by one of the companies that manufactures these machines and does analyses on a contract basis. This scientist left the session believing that although the company's primary clientele for contracted services was industrial, the concern would "do analysis for researchers for free, if . . . you're just a university researcher looking to publish it." He was excited and soon wrote the company explaining the lab's project and requested the free service. He hoped they would analyze in the neighborhood of a hundred isolates and offered to pay the cost of the analysis, which he thought would be around two dollars per isolate.

The company said they would be unable to undertake such a large analysis at cost, but offered to do four strains for free. The Handelsman lab got back the results of this analysis and it looked as if the "correlation was still holding up." That is, the analysis suggested that *B. cereus* isolates that produce zwittermicin A had a particular fatty acid profile. Handelsman and the student decided that it would be worth testing additional strains, and they sent twenty strains to the company, agreeing to pay the so-called researchers' rate of twenty-three dollars a strain. In this case, as in the others, UW85, the *Bacillus cereus* strain about which the lab knew the most, was sent as a control.

This larger run was the fourth time the lab had had fatty acid analysis done. In the first three efforts, the results suggested that the control, UW85, came out with the same profile. More precisely, when the analysis of the fatty acid data is done, samples are compared using a method called cluster analysis. The results provide a graphic representation that could be called a "relatedness tree." The first three times UW85 was tested it came out on the "same branch of the tree." On this larger run, UW85 came out on a different branch of the tree. As the scientist overseeing the project put it: "So—my control didn't work because it should be the same, identical."

Without an effective control, lab members could not have confidence in the rest of the data. The student overseeing the project felt that the company may have made an error in its testing. He phoned and spoke to a staff scientist to explain the problem. He wondered whether perhaps he had used a different mathematical analysis for clustering than had been used on previous runs, or whether something might have been done differently in the testing. He noted that at the seminar he attended the presenter indicated that the technology has a very high degree of reliability. According to the Handelsman lab graduate student, the staff scientist simply repeated that her company could not have done anything wrong.[12] The Handelsman lab worker found this infuriating because "you can never say that. . . . [T]here could have been a mistake on either end." The Handelsman lab researcher, however, had been exceedingly careful and had precise copies of all the material he sent off. He imagined the problem was at the company's end.

He followed up with a letter to the company in which he suggested, "Given the distance between the same strain run twice it looks as though either something was amiss during one of the runs or fatty acid analysis isn't as reliable as I thought it was." The Handelsman lab worker continued, "If the former is the case, I think you should redo the analysis. If the latter is the case, I suppose its [*sic*] important that this is reported to people so that they can be more cautious about what claims are made based on this sort of analysis" (lab correspondence to company). This was the beginning of a dispute between the Handelsman lab and the company—a disagreement that is ultimately based in the difference between the culture of academic science and that of industry and the market (see Dubinskas 1988)—and was affected by asymmetries of resources and power.

It was some time before the Handelsman lab received a response from the head of the company. The letter from the official stressed the lab scientist's request to have the analysis redone "free of charge." The official pointed out that the Handelsman lab had been given a cost break the first time around. He noted that the savings to the lab amounted to "$494 we gave away because I assumed that as a plant pathologist you had little money and were a nice guy." The company leader drew attention further to the Handelsman scientist's assertion that if the system is not as reliable as he was led to believe it was, this should be reported in the scientific literature. The letter concludes:

> It seems to me that you are trying to extort additional free analyses from [the company] . . . by threatening to report that the claims made about this type analysis are false. Therefore, I shall respond in kind and tell you that if you publish such erroneous and untrue statements that it will be seriously damaging financially to my companies and I will take the necessary legal steps to obtain redress from you and from your employer. (company correspondence to lab)

The head of the firm, as well as company employees involved in the case, believed the analysis had been done correctly and that the problem was on the lab's end.

We find here a complex resource dependence relationship combined with a clash of cultures. This is a relationship of resource dependence at two levels. First, it is a dependency relationship economically. The Handelsman lab cannot afford to buy the technology to do the tests they want. Consequently, they must rely on a company that does these tests for profit.[13] At a second level, since there are few companies doing this test commercially, the Handelsman lab had limited options. These two dimensions of resource dependence are tightly linked. Even with the possibility that other facilities could do their work, the Handelsman lab must consider costs, and this company had given the lab a discounted price.

Both the Handelsman lab scientist leading this project and the leader of this company are concerned about their views of truth, their own reputations, and the implications that claims of falsity would have for their reputations. I do not read extortion into the letter sent to the company on behalf of the Handelsman lab. Instead, the student who wrote the letter uses the language of science—of reliability and caution. The explicit statement he made is concerned with facilitating the collective knowledge production project of the community of scholars of which he is a part. He tells the company leader that if the test is not as reliable as is claimed, the claims should be modified. The smooth functioning of the scientific enterprise depends on trust (Shapin 1994), and researchers need to be able to trust that tests that are claimed reliable will, indeed, be reliable.[14] In this case, the problem of reliability is complicated further because the lab cannot afford to undertake the tests itself. The lab cannot purchase the equipment and thus must depend on and trust the company.

The reputation of the Handelsman lab depends on accuracy of research results, and this is no less the case for the company. However, in academic science there is a direct relationship between accuracy of research and reputation. Scientists are assessed by their colleagues on the basis of the peer-recognized quality of their research (Brooks 1993, 217; Slaughter and Leslie 1997, 116). In science-based industry, reputation is a mediating factor between accuracy and profit. Accurate results are necessary to retain clients and consequently profit.

The cultural differences between the Handelsman lab and the company, as well as the power asymmetry between the two are further highlighted by the threat of legal action. With this threat, the lab is left with little choice. Although lab members can informally warn colleagues away from this company, they cannot pressure the company to report any modification in the reliability claims they make about the technology. They have no power to do so. Indeed, the lab cannot publicly question the reliability of the company's tests, unless it has the wherewithal to contest the company's claims in the realm of law—in the courts. Here, the rules are different than in the realm of science, and the cost is likely to be prohibitive. In sum, a dispute about the source of inaccuracy was not contested with the traditional tools of science—"evidence" and argument.[15] An asymmetry of financial capacity did not facilitate agreement, but instead coerced silence.

Here is a case where Fujimura's cooperation between social worlds broke down. But Bruno Latour's concept of the "obligatory passage point" might be applied here. Certainly at some level fatty acid analysis became an obligatory passage point for the Handelsman lab. This concept would point us to the lab's dependence on the tool and *its relative lack of power regarding it*. However, the concept tells us nothing about the system or larger environment in which fatty acid analysis and the lab's relationship to it is embedded. In my view, the key fact here is that the Handelsman lab is a relatively low-budget university lab. The equipment for fatty acid analysis is expensive and the lab cannot afford to buy it. This makes the lab dependent not only on the technology, but on those who possess the equipment to perform it. It is important in this case that it is for-profit companies that provide the service, and they do not normally offer their services at cost. Additionally, although the Handelsman lab worker at the center of this conflict is interested in opening up this black box, it is not the cost to the relatively small scientific community that prohibits this challenge by the lab, but rather the threat of legal challenge by an economically robust entity that induces lab workers to make no public criticism of the company that provided it with fatty acid analysis services. Finally, this marks the contrast between the actor network approach and my emphasis on structure. Unlike what we would notice from an actor network perspective, in this case we see a particular structure that constitutes a distinctive set of resource distributions, capacities, and incapacities; it defines variations in opportunity and constraint by structural location.

Tic-*Taq*-Toe: A Black Box Makes It Hard to See

If the case of fatty acid analysis points to one way in which big biology and the reliance on for-profit entities can affect laboratory practice, another episode involving a collaboration between the Handelsman and Goodman labs and a polymerase called *Taq*, which is used in the polymerase chain reaction (PCR), points to another. As I explained in chapter 1, when one heats the double helix structure of a piece of DNA, PCR separates the twisted strands into two individual strands. The primers—short single stranded molecules—bind to the ends of the now-single strands of DNA, and nucleic acids fill in the remaining positions in the sequence, creating a new double helix. This process is repeated, permitting the geometrical amplification of the target DNA fragment.

The PCR process was initially developed by Kary Mullis, while working at the Cetus Corporation, a California-based biotechnology company (Rabinow 1996). His findings were reported in *Science* in 1985. PCR was patented in 1987 and commercialized by Cetus in 1988. Called "one of the most important and powerful tools in molecular biology," between its initial development and the fall of 1994, research for over 25,000 scientific articles relied on PCR (Alexander 1994). According to a 1993 report, more than 70 percent of molecular biology researchers utilize PCR as a tool in basic research (*Business Wire* 1993). Mullis won the 1993 Nobel prize in chemistry for his work.

During the initial development of PCR, scientists had to introduce fresh polymerase during each amplification cycle because the heat required to separate the strands of DNA inactivated the polymerase. This laborious step was eliminated when scientists from Cetus determined that using an enzyme from a bacteria that grows in the boiling waters of Yellowstone National Park in Wyoming—*Thermus aquaticus*—would eliminate the need to introduce new material during each cycle. The polymerase *Taq* survived the repeated heating and cooling.

In the microbial ecology project undertaken jointly by the Handelsman and Goodman labs, PCR plays a central role. This project involves efforts to assess the diversity of bacteria in microbial communities in soil. Researchers use what is called a molecular phylogenetic approach. Such an orientation involves isolating "total DNA" from soil samples. This strategy has the advantage of avoiding the necessity of culturing bacteria from soil in order to catalog it. Culturing—growing bacteria on media in incubators—can only be accomplished successfully for a small portion of all soil bacteria. Although distinct from work on UW85, the microbial ecology project is related to the UW85 investigation insofar as this assessment could lead to the discovery of additional novel antibiotics and perhaps new biocontrol agents.

Researchers working on the project bypass culturing by extracting genetic material directly from the soil and using PCR to copy or "amplify" specific pieces of DNA known as the 16S ribosomal RNA gene. This specific gene is useful because all organisms have "very high identity" in terms of it. As a result, small differences between these genes among organisms provide a signature for each individual and permit researchers to categorize previously undiscovered soil bacteria.

Investigators on the microbial ecology project rely on the *Taq* polymerase in their amplification efforts. In this particular project, the assumption that commercially available *Taq* was appropriate turned out to be a mistake. In initiating this work, the postdoctoral researcher in charge of amplifying 16S genes from soil samples set up a negative control—a tube containing the various chemicals used in PCR—including *Taq*—and water, but without any DNA from soil. After amplification, he ran his DNA samples and his controls on an electrophoresis gel to assess the success of amplification. Over several months, the postdoc successfully amplified DNA from his soil samples, but there was a discouraging problem as well: the negative control appeared to contain the 16S gene. The scientist considered a variety of sources of contamination, including the possibility that the water contained some microorganism with the 16S gene.

For some time, lab workers focused on eliminating contamination from the water, using materials that would rid the liquid of DNA. After several months, lab leader Goodman happened to talk to a researcher who also worked with the 16S gene and learned that contamination through *Taq* is a common problem in work on this particular gene. *Taq* is commercially "grown" in the common gut bacteria called *Escherichia coli*. *E. coli* contains a version of the 16S gene. For most researchers who use *Taq* in PCR the presence of a small quantity of *E. coli* DNA is not a problem because their practices would not lead to the *E. coli* DNA being amplified.

Scientists in the Handelsman and Goodman labs are quite aware of the "contingency of the local." That is, they are acutely conscious of the fact that even when efforts are made to strictly control experimental practice, variability can never be entirely controlled. However, because many research tools have become black boxes (see Fujimura 1992; Latour 1987), scientists typically assume they work properly (Fujimura 1996, 77). Indeed, in her experience, Fujimura found that scientists working with recombinant DNA technologies "attributed failures to inattention to detail or contaminated materials, and not to problems with the techniques" (1996, 109). Fujimura suggests further that when a standard technology reliably reproduces results across situations, it is unlikely that they will be opened up and tinkered with. This is "not commonly done after a set of protocols has been widely adopted" (1996, 105). In short, when an experiment using a standard technology or technique goes awry, it is the local (laboratory conditions or a scientist's failure) that is likely to be scrutinized first.

To participate in the contemporary molecular biology field, university laboratories must rely on materials supplied by commercial concerns. These materials are standardized: they are created to suit a wide market of laboratories, not the particular needs of individual labs. The configuration of these materials can be taken for granted—they are black boxes—and the researchers' state of necessary ignorance can adversely affect laboratory practice. Here, the context in which a laboratory exists today—big biology in the age of commercially produced standardized research tools—can be said to shape laboratory practice.

This case illustrates the existence of a power dynamic between the research materials supply industry and university laboratories, which can be understood on two levels. First, the *Taq* polymerase comes in a limited number of variations from a limited number of supply houses, and to participate effectively in the world of molecular biological research, university labs are dependent on these companies for *Taq* and similar standardized resources (Pfeffer and Salancik 1978). It would be crippling if university laboratories found it necessary to customize each of the tools and materials they use daily. If they are doing work in which the standardized tool is appropriate, standardization works to their advantage. If, however, a lab's needs do not match the market demand that companies respond to in their products and services, the lab is left with the costly (and sometimes impossible) task of custom making its tools.[16] Second, because to work effectively researchers must trust the efficacy (Shapin 1994) of commercially supplied materials, they are disinclined to scrutinize the contents. In accepting the role of suppliers to define the content of these tools—these black boxes—they are dependent on the suppliers. Thus, the supply industry *indirectly* defines the practical mechanics of laboratory work.

Kit Science and More: The Trouble with Troubleshooting

During my time in the laboratory, I heard "kits" both savaged and lauded. Kit producers were accused of providing mislabeled or poor quality material, and they were credited with facilitating completion of experimental work on time. Kits are packages of research materials produced by biological supply companies. There are an array of kits on the market. There are kits for cloning DNA and for purifying genetic material, kits to "facilitate gene discovery" (Intirogen 1997, ix), and to aid in gene library construction. The character of these kits varies, but their general purpose is to increase efficiency and reliability by providing all (or many) of the materials necessary to undertake a test or procedure in a standardized form.

One issue specifically raised by Handelsman and another scientist in the lab concerned the effect these tools can have on scientific training and troubleshooting skills. When you are given a recipe and a set of premeasured ingredients to do a procedure, there is a tendency to ignore the precise workings of these "black boxes" and, in theory at least, this could cause the scientists who use them to be less knowledgeable or less able to troubleshoot than scientists educated in the days before widespread kit use. One staff scientist in the Handelsman lab said, "If you don't understand what the kit is, you can get into trouble." A graduate student's reflection reinforces this point: "I know a lot of people when they do biology, like using kits, mix 'solution a' with 'solution b' . . . and really don't know what they are actually doing." In light of these comments, I set out to explore this issue further. I interviewed many of the students and others in the lab about their sentiments about kits. I also asked Handelsman and a staff scientist to explore with me at greater length the effect of kits on training.

The staff scientist described for me the possible problems that might arise if a student scientist were to use a kit for doing a procedure known as a Southern blot. A Southern, as the technique is referred to in the lab, is a means to "probe" for specific bits of DNA, a method for identifying a specific piece of DNA from a mixture of pieces of DNA. It involves running an electrical current through a "gel" as a means for getting pieces of DNA of different sizes to "migrate" to different points on the gel. The gel, in other words, is used to separate the mixture of pieces of DNA by size. Once the gel has been run, the DNA is transferred to a hydra-cellulose membrane. Then some DNA with characteristics known to the investigator is used to learn more about DNA about which she has limited knowledge.

The staff scientist described the kits used in the lab with non-radioactive probes as working "beautifully." Most of the chemicals come premade. Most importantly, the materials for labeling the probes are provided. The researcher can then make a probe to search for a particular region of a gene. However, according to the researcher, "how big your probe is is one of . . . [the] variables that affects how the kit works." And the nature of the probe constructed affects the results obtained. If a novice simply follows the "beautiful instructions" without understanding the limitations of the kit, the results will be affected. Indeed, the kit might not work at all. The staff scientist tells of one instance:

> There was a student who got to a certain point who'd sort of done everything up to the point where the blot was ready to probe. Basically . . . [the student] hadn't done [all of the steps in the procedure]. . . . [The student had] done the probing, but after you probe . . . you have to do a detection. . . . Well—the student had done the probing and . . . said "I don't have any bands. It didn't work." Well—it wasn't that it didn't work; it wasn't done.

Concluding this story, the staff researcher commented that her assumption is that if "someone is going to use a kit they understand how it works. . . . [I]f you have to do something from scratch, you have to be aware of just a whole range of other nuances in the protocol that with a kit you" might ignore.

Of course, one can also succeed in a case like this. Most of the people in the lab I spoke to about kits remarked that the best kits have detailed and comprehensive instructions. Consequently, "if you follow the instructions, you don't have to understand what it does. It will work" (staff scientist interview). But in this instance, a student can complete the experiment and not understand why it worked. According to the staff scientist, the reliability of kits like the Southern blot kit she described can lead not only to ignorance among apprentice scientists, but also to carelessness among more advanced researchers. They know that the kit works and consequently do not run the pretests provided with the kits prior to doing their experiments.

The issue of standardization and student knowledge came up for Handelsman when she was participating in a dissertation defense after I had left the lab. Handelsman noticed that the student had neglected to italicize the first three letters of the names of the restriction enzymes she used in her work. Bacteria have Latin names and the practice is to italicize them. Since restriction enzymes come from bacteria, the first three letters of their names are also italicized. For example, in the case of the restriction enzyme known as *Hin*DIII, the letters H-I-N are short for *Haemophiles influenza*. Similarly, the restriction enzyme *Eco*RI comes from the bacteria *E. coli*. According to Handelsman, bacteriologists rarely fail to italicize restriction enzymes because they know where they come from. On the other hand, virologists, like the student in question, are less likely to have worked with the bacteria from which restriction enzymes come. Because these tools are typically purchased from biological materials suppliers, virology students are less likely to be aware of the bacterial source of the restriction enzymes with which they work. Indeed, when I was with Handelsman sometime after the dissertation defense, the same virology student walked by the office where we were talking. Handelsman asked her if she knew where *Hin*DIII came from. The student said she did not. Handelsman provided this analysis of the situation:

> Now I can't believe that if she had to purify her own, as people did in the lab in the late seventies, that she would have not known. She would have been growing up a culture of *Haemophiles* and isolating the enzymes. So that's a level, that's not kits, but is purchased versus made in-house kinds of technology and that just struck me a lot like the sort of the food in the grocery [store]. That this is

> just something you call up Promega, and you buy it from them. She had no idea that this is an enzyme from an organism that plays a role in that organism's life, and maybe that is a sort of divorcing from what you're really using this for. . . . [W]hat you use it for and where it comes from are connected at some degree. For example, if you are trying to troubleshoot, it might help to remember what the middle of the *E. coli* cell looks like. . . . If you're enzyme isn't working maybe thinking about where the enzyme came from and the conditions under which it normally does work would help.[17]

Of course, this is precisely the point of standardization and commercial production of research tools. As one analyst describes restriction enzymes in particular, they come "to contemporary users as . . . black box[es] embodying the cumulative work of many scientists on the enzyme's activities. Each enzyme is ready-made for immediate use, for specific purposes and under specific conditions" (Fujimura 1996, 81). As molecular biology came of age, the standardization of procedures and the designing of black boxes that even the uninitiated could use was precisely the objective of many researchers. An early laboratory manual for cloning produced at Cold Spring Harbor Laboratory "was actually written in a way that allowed people from virtually every field to come in—with no understanding of lambda, or no understanding of bacteria—to read the manual and do it" (quoted in Fujimura 1996, 84).

Handelsman indicated that similar problems of ignorance have arisen when students work with a National Institutes of Health database used to look for DNA sequence similarity.[18] According to Handelsman,

> The programs work by pretty clear parameters in terms of aligning your sequence with what's there and asking for matches, and that may sound like a very simple thing, but when you think about the computational issues involved obviously if you have two truly identical things, if you have a good program, it should be a pretty simple thing to line them up, but what you're looking for is not identity, but similarity, and so what do you do with gaps? Like what if one gene has this big area in it, this big region, that your gene doesn't have? What do you do with that? Do

> you . . . just sort of ignore it, kind of loop it out and then align the parts that are similar? What if one gene is much longer at one end and you give them the same start site? Well-they're never going to align right then because the similar regions in one are going to be much further down the road than in the other.

In these situations, Handelsman has found that students sometimes come to her satisfied that they have found a match, but upon further discussion it becomes clear to her that the students did not entirely understand the algorithms underlying the technology's analysis. They had not opened this black box. Consequently, they make faulty assumptions about how to align their gene sequence with the data provided by the database, and may arrive at inappropriate conclusions about the nature of similarity. With this technology the matter is complicated further by problems of specialization. A clear understanding of this technology might require the aid of a computer scientist, but "there are so few computer scientists to actually sit down and explain how you might come up with an anomalous results from an algorithm that is built to do one [particular] thing" (Handelsman interview).

Handelsman's concern about gaps in the education of current graduate students caused by the commercialization and standardization of research tools is echoed by biologist Harry Rubin, who suggests that "older, slower methods often force scientists to think through their problems more carefully" (quoted in Fujimura 1996, 112). However, Handelsman also notes that in theory commercially produced kits can be excellent learning tools because they often provide detailed explanations of how they work.[19] Additionally, some analysts point out that the increasing use of standardized tools has not led to a systematic degradation of the skills of apprentice scientists. These students may not learn things that they would have learned decades earlier—and will not acquire the same skills, but they obtain other, different knowledge and new skills (Gaudilliere and Lowy 1998a, 11). Handelsman and several of her students pointed out as well that the problem is in the black-box nature of these technologies, not in the fact that they are commercially produced. This fact is illustrated both by the way in which the Cold Spring Harbor manual makes cloning possible for novices, and the publicly funded database for assessing genetic identity described above. Though it is not a commercial service, it can permit student researchers to proceed without knowing how it works. At the same time, in an environment in which a great deal of standardization has occurred and where commercial opportunities often tend to propel standardization, data demands have increased. Handelsman noted, repeating the comment of one of her students, "the demands for what it take to make a paper are so great compared to what they were even five years ago that if we had to do it the old way, you would take ten years to get a Ph.D.—to get enough data to publish your paper." Thus, even though careful researchers who receive their degrees today may be more knowledgeable than their predecessors, the pressures of fast-paced science may compel the novice as well as the expert to forego close examination of black boxes, commercially produced or otherwise. Sometimes this will not matter. Other times it will. In any case, today's scientists' work—their outlook and daily practice—has been shaped by the widespread availability of standardized research tools.

Conclusion

It is certainly true that the standardization and accompanying commercialization of the research tools used in the life sciences in university labs have made research more efficient. These technologies clearly permit scientists to study phenomena that were previously inaccessible in ways that were previously impossible. Using fatty acid analysis, Handelsman lab members can now locate zwittermicin A producing *B. cereus* strains more effectively and more efficiently. Handelsman researchers working on the direct extraction of DNA from soil published a major paper on microbial ecology; the research for the paper would have been impossible without PCR technology (Bintrim et al. 1997). Handelsman regularly sees her students graduate as talented, knowledgeable Ph.D.s.

There is little doubt that it makes sense to talk about how inscription devices make fact construction possible (Latour and Woolgar 1986 [1979]), and to consider how standardized packages permit cooperation across social worlds (Fujimura 1987;

1996). Clearly, Knorr Cetina is correct to point out that data banks, standardized packages, and automated technologies free molecular biologists "from routine tasks and allow them to spend more time explor[ing] 'more interesting' goals and new techniques" (1999, 84). At the same time, university biologists work in a world that is not entirely of their own making. It is a realm that has been structured in significant ways by the commercialization and the standardization of research tools in the biological sciences. These tools are often expensive, and to participate effectively in the community of scholars doing research in their sub-field of biology, academic laboratories—like the Handelsman lab—must purchase investigative tools and materials or contract for research services. Researchers are frequently in a relationship of resource dependence with their suppliers. They are not able to produce certain research tools themselves and the number of suppliers to which the lab can turn is limited. In cases of disagreement between firms and scientists, the scientists are often powerless to challenge the former. They must make do with what is available. Furthermore, in the fast-paced world of the life sciences, standardized—often commercially produced—research tools are commonly taken for granted, and accepted as pure, appropriate, and effective. As a consequence, researchers may not examine the black boxes they use when they should be opened; they then fail to learn what their predecessors learned through necessity.

It is not my purpose here to make a moral assessment of scientists' increasing reliance on commercially produced and standardized research equipment and other materials. I do not wish to ignore the breakthroughs made possible by these tools, nor do I wish to minimize the drudgery that can now be avoided. I mean only to point out that there are ways in which these developments shape the practice of academic biology that are not always overt, and to illustrate that these effects are often entangled with issues of asymmetrical power and cultural conflict.

5

Owning Science

Intellectual Property and Laboratory Life

A selection of articles appearing in *Science* magazine in the late 1990s points to the growing concern about intellectual property in the increasingly commercialized world of the biological sciences. One article discusses a debate over patenting human gene fragments (*Science* 1997b, 187). Another talks of the industry-supported work of academic "genetic prospectors," researchers who are "scouring odd corners of the world for families whose DNA is likely to carry interesting genes" (Marshall 1997, 565).[1] These scientists and their collaborators intend to patent the "interesting" genes they discover. Still another pair of articles considers a debate over ownership and control of data on the genetic composition of the entire human population of Iceland. The company that would undertake this data collection wants to be able to sell access to it. Some scientists worry that the company will restrict access when restriction rather than access may be in its economic interest (Enserink 1998a and 1998b).[2]

The trend is clear: the products of biological research from both inside and outside the university are increasingly treated as private property and protected by U.S. patent law. With federal support for academic science limping behind demand, this trend could be a boon. Companies interested in deriving profit from the revolution in biology often turn to university labs. In exchange for research support, companies can gain access to proprietary inventions. At the same time, some researchers and policymakers express concern that commitments to intellectual property protection in university settings can undermine traditions of free intellectual exchange. And recent developments illustrate how the patenting of research tools could hinder scientific investigation.

The literature that explicitly addresses intellectual property matters and academic science is extensive. As I noted in chapter 2, a central concern of this work is the increasing tendency of academic biologists—especially those scientists who have contractual relationships with commercial entities—to seek patent protection of their work, and the impact that action has or will have on the traditional commitment in academe to the free flow of information and research materials (see AAUP 1983; Kenney 1986; Shenk 1999).[3] The primary debates in science studies give little attention to intellectual property matters. In *Laboratory Life,* Latour and Woolgar tell us that the laboratory they studied had "many connections with clinicians and industry through patents" (1986 [1979], 182), but they provide no analysis of the role of intellectual property considerations in the fact construction process or in the workings of the laboratory. Similarly, Knorr Cetina mentions intellectual property matters, but provides no analysis of the existence of relationships with commercial concerns and applied interests among scientists in the laboratory she studied for her 1981 ethnography. Similarly, in her more recent comparative laboratory study, Knorr Cetina mentions that research materials can be patented (1999, 154) but provides no sustained discussion of this state of affairs.

In the 1990s several articles on patenting and the biological sciences that explicitly deploy the conceptual tools currently used in science studies were published. In one paper, Alberto Cambrosio and his colleagues (1990) investigate patent litigation involving a dispute over immunoassays between two biotechnology firms. The essay focuses on "rhetoric and representations" (Cambrosio, Keating, and MacKenzie 1990, 277), exploring "how things perceived as naturally given are constructed and represented" (1990, 277). The authors' concern is mainly with how categories are constructed, as well as how they are blurred in the context of the dispute.

Using a related approach, Greg Myers (1995) follows two academic scientists as they navigate the world of patenting. Myers explores the differences between patents and scientific articles in the ways they make claims, how they relate to other texts, and how they narrate future action (1995, 58). Myers draws on the concept of enrollment used by actor-network analysts, and tries to show how the scientists he follows configure actors ranging from experimental animals to lawyers and companies (1995, 60). He argues that when a patent succeeds, it creates the social world around it (1995, 99).[4]

In this chapter, I provide a different perspective on intellectual property protection and the academy than that provided by much of the current science studies scholarship and science policy literature. Whereas ethnographies like Latour and Woolgar's, and Knorr Cetina's barely mention intellectual property matters, I believe that a serious study of contemporary academic biology demands a substantial analysis of intellectual property issues. Unlike scholars of university-industry relations (UIRs), I contend that more attention needs to be paid to the structure of constraints inherent in the intellectual property regime in which academic scientists in the United States operate than to how patent-related restrictions of information result from relationships with for-profit entities. I agree with the scholarship of Cambrosio and his colleagues and of Myers that we must be attentive to rhetoric in patent disputes, and that, as I say in chapter 2, actors engage in constructive activities. At the same time, intellectual property protection takes place in an already constructed world in which rules of play, as well as the variety and value of resources are well established. These rules and resources shape the space in which actors have to maneuver and determine the kinds of strategies most likely to succeed and fail.[5]

In this chapter, I focus particularly on the ways in which the intellectual property regime in the United States affects workers in the Handelsman laboratory and other university scientists. I use the phrase "intellectual property regime" to mean laws, court decisions, and associated institutions established for protecting and exploiting intellectual property, as well as commonly held attitudes concerning intellectual property. I consider three ways in which the practices of academic biologists, and members of the Handelsman lab in particular, are shaped by this intellectual property regime. First, I explore how corporate patenting of research tools can affect the everyday practices of scientists. Second, I consider traditional arguments concerning the dangers of patenting in a university environment; I contrast the Handelsman laboratory's own experience in encountering barriers to the free flow of information and research materials with the overlooked issue of the potent influence of commonly held attitudes about intellectual property protection. Finally, I analyze how the Handelsman lab has been affected by its relationship with the University of Wisconsin's patent agent, the Wisconsin Alumni Research Foundation (WARF).

An examination of these three aspects of intellectual property suggests that the typical concerns people have with increased secrecy in academic biology fail to take into account more complicated and less easily resolved issues: first, who controls research, and second, how and what is happening not just to the culture of the academy, but to the knowledge commons. These issues turn attention from the direct impact of specific UIRs and illuminate the indirect and pervasive influences of the intellectual property regime on university biology in the United States. However, before turning to these issues, I provide a brief discussion of patenting and the revolution in molecular biology.

DNA and Bacteria as Property

The onset of the intellectual property revolution in academic biology coincided with the development of recombinant DNA techniques. Before the mid-1970s, virtually all research in molecular biology was funded by the federal government or private foundations. In this context, rights to patentable inventions—of which there were few—went to sponsoring organizations. During the 1970s, however, a collaboration between Herbert Boyer at the University of California–San Francisco and Stanley Cohen at Stanford changed everything.

In biology attention was already focused on the possibilities of altering genes to change the proteins they produce. Cohen and Boyer made this a reality. Boyer figured out how to use a so-called restriction enzyme to splice double stranded DNA at a fixed spot. Cohen and Boyer were then able to cut a plasmid—a ring of DNA that sits outside the main chromosome in bacteria—and insert DNA from another organism, using the snipping power of restriction enzymes, into the plasmid (Teitelman 1989,

18, 19). This was the beginning of recombinant DNA (rDNA) technology; in November of 1974, Stanford University filed for patents on the basic rDNA procedures that Cohen and Boyer developed and on the plasmid, pSC101, used in the procedures. After this point, private appropriation of patentable results in molecular biology became increasingly common (Wright 1986).

If Cohen's and Boyer's efforts laid the groundwork for the commodification of biologicals, the environment in which intellectual property considerations were undertaken was changed significantly by several judicial and policy actions. On the judicial front, 1980 was an important year. In that year, the Supreme Court ruled in the case of *Diamond v. Chakrabarty* and concluded that living organisms could be patented. Some eight years earlier, Ananda Chakrabarty, a microbiologist at General Electric, applied for a patent on a microorganism he genetically engineered to break down most of the hydrocarbons in petroleum and that could potentially be used to clean up oil spills. The "bugs" would "eat" the oil. His application was rejected on the ground that U.S. law did not permit patent protection of living organisms, except for plants suitable for such protection under the Plant Patent Act. The decision of the U.S. Patent and Trademark Office on the Chakrabarty application was reversed by the U.S. Court of Appeals. Unhappy with that reversal, the Office took its claim to the Supreme Court in the name of the Commissioner of Patents and Trademarks, Sidney Diamond. There, the Appeals Court's decision was upheld. The microorganism for which Chakrabarty sought protection was not a product of nature, the Court ruled, but of a scientist's creativity, imagination, and initiative.

Another significant step in the commercialization of biology occurred in 1985 when the *Chakrabarty* decision was broadened by the U.S. Patent and Trademark Board of Appeals, when Kenneth Hibberd appealed the Patent and Trademark Office's (PTO) rejection of his application for a patent on a variety of genetically engineered corn he had developed while working for a Minnesota-based biotechnology company. In *Ex parte Hibberd,* the Board ruled that any type of plant could, in principle, be patented. This determination was further extended in 1987; the PTO ruled then that all living animals could be patented (Kloppenburg 1988; Bugos and Kevles 1992). This group of decisions defines the contours of the current legal framework for the protection of intellectual property in the life sciences in the United States.

Congressional action in 1980 also served to facilitate the commercialization of the life sciences. The Bayh-Dole Act of that year stated the objectives of Congress clearly: "to promote collaboration between commercial concerns and nonprofit organizations, including universities" (quoted in Slaughter and Rhodes 1996, 317, 318). The act permitted universities and small businesses to retain title to inventions produced during research undertaken with federal support. Prior to the Act's passage, "universities were able to secure patents on federally funded research only when the federal government, through a long and cumbersome application process, granted special approval" (Slaughter and Leslie 1997, 46).

As developments in research proceeded, commercial considerations prompted the biotechnology industry as well as some academic scientists to seek further broadening of what could be patented. In February of 1997, an official from the Patent and Trademark Office announced at a scientific meeting that the Office would henceforth grant patents on short stretches of DNA called "expressed sequence tags," or ESTs. The decision to permit the patenting of ESTs was based on their utility as research tools. ESTs can be used as probes to identify specific DNA sequences. While applauded by some, this decision worried others, including officials at the National Institutes of Health, who believed that this policy would lead to restrictions in the flow of genetic data in the scientific community (*Science* 1997b, 187; *Science* 1997c, 41). In January of 2001, the Patent and Trademark Office made a move that bucked the trend toward broadening the scope of the commodification of scientific information when it raised the bar on what counts as utility in an invention (Reuters 2001; U.S. Patent and Trademark Office 2001). This new requirement could lead many EST patent applications to be rejected (Enserink 2000). Despite this apparent reversal of a trend, in the years ahead, the scope for intellectual property protection in biological materials is likely to expand rather than contract.

Property Rights and Research Tools

Academic administrators, scholars, and policy analysts have devoted a great deal of attention to how intellectual property considerations in academic science have influenced the character of scholarly communication, and also, more narrowly, the free flow of scientific information (see Giamatti 1982; Kenney 1986; Shenk 1999; Lacy 2000).[6] However, the increasing commodification of research tools used in the basic biological sciences raises a series of different questions about who controls these technologies and substances and what the implications may be. In this section, I consider four instances in which concern about intellectual property considerations regarding pivotal research tools could significantly influence the daily practices of academic scientists.

The first example involves a polymerase, *Taq*. Beyond its function as a black box, as I discussed in the previous chapter, *Taq* is intellectual property; *Taq*'s proprietary character can potentially limit a university lab's freedom of movement. *Taq* is used in the Handelsman laboratory and widely used in molecular biology laboratories elsewhere. As I discussed, it was used in the Handelsman and Goodman labs' joint microbial ecology project, and I used the substance regularly in my efforts to locate a gene in bacterial samples that codes for resistance to the Handelsman lab's antibiotic, zwittermicin A.

What does it mean when the research tools commonly used in academic laboratories are proprietary, that is, when these tools are owned by companies interested in making a profit from their sale and use? In theory, this should make very little difference because legal tradition in the United States permits researchers interested in fundamental investigation to circumvent patents for the purposes of study. Basic research relies on a principle known as the "experimental use exemption."[7] According to this concept, scientists engaged in "basic science" are allowed to use patented products, processes and other publicly disclosed intellectual property for research purposes without license or explicit permission. These investigators could typically avoid restrictions and fees associated with licensed use of patented inventions. But this research exemption was recently threatened by a corporation interested in protecting its investment in *Taq*, a proprietary research tool.

The importance of the proprietary status of *Taq* became clear to me as I was led outside the lab by an article discussed at a weekly meeting of "journal club"—the gathering held jointly by the Handelsman and Goodman labs to discuss recently published scientific papers. Attention turned to a news item that appeared in *Nature* (1995) that described a patent controversy surrounding the *Taq* polymerase. The basic controversy pits the Wisconsin-based biotechnology company, Promega, against Hoffmann-La Roche (now the Roche Group), the multinational healthcare corporation; it has implications that could reverberate well beyond this corporate conflict.

Taq is an extremophile enzyme or extremeozyme, an enzyme that comes from a microbe that lives in extreme conditions. In this case, as I discussed in the previous chapter, that environment is the thermal pools of Yellowstone National Park in Wyoming. *Taq* comes from the bacteria *Thermus aquaticus,* which was discovered, interestingly enough, by a University of Wisconsin scientist, Thomas Brock, in 1965.[8] Several decades later, Cetus Corporation patented *Taq* after company researchers determined it could vastly improve the efficiency of the Polymerase Chain Reaction (PCR).

Roche acquired the *Taq* and PCR patents from Cetus in 1991. The company paid $330 million to acquire the patent for *Taq* alone (Reuters, 1993). At the time, Promega had a license from Cetus to sell the *Taq* polymerase as a generic enzyme, but not as a tool for use in PCR. Roche believed that Promega was violating its license by selling *Taq* to researchers specifically for PCR, and thereby reducing Roche's profits. A complex lawsuit developed over the status of *Taq,*[9] and as part of its efforts to dam the leak of its profits, in May of 1995, Roche named 40 universities and government laboratories in the United States, and 200 individual academic researchers as patent violators (BBI Newsletter 1995).[10] These scientists used *Taq* purchased from Promega for running PCR. But because many of these university scientists are engaged in basic research, Roche's claim amounts to a direct challenge to the experimental use exemption.[11]

How do the Handelsman and Goodman laboratories fit into this controversy? From the *Nature* article, the discussion in the lab meeting turned to consideration of the possibility of manufacturing *Taq* in-house. Practical details were the focus of discussion. Might the two labs hire a technician to produce purified *Taq* for the research undertaken in both labs? If they did so, would this be more cost effective than purchasing *Taq* commercially?

After the meeting, Handelsman told me that there were two reasons she and Goodman would consider producing their own *Taq*. The first of these brings us back to the discussion of the microbial ecology project in chapter 3. Because commercially produced *Taq* is not "clean enough" for the purposes of this project, Handelsman and Goodman thought it worth considering manufacturing the polymerase themselves. But in addition Handelsman said that "when the patent wars started it looked like there were all these controversies that kind of made us think about it on another level . . . [:] the price is just going to keep going up and maybe we'd save a lot of money if we did our own as well" (interview, April 7, 1995).[12] Indeed, if Roche were able to alter the licensing situation and the royalties it demands, the price of *Taq* could, indeed, go up. In this context, a scientist writing to *Science* complained that the cost of purchasing PCR products, as compared to producing them in-house, would be six times the annual research budget for the scientist's lab (Packer and Webster 1997, 53).

PCR and *Taq* were developed in an environment that bears very little resemblance to the romantic vision of the scientific community as a world of Mertonian communism (1973 [1942]) and perfect information. *Taq* may be a standardized technology that permits cooperation across diverse social worlds, as Fujimura (1996) might put it, but it also is a profit center. It is a standardized package—a combination of materials and theory—that embodies, or perhaps masks, a particular set of property relations. *Taq*, an essential research material, is the property of a company interested in making a profit. Through ownership of the patent on *Taq*, Roche has a monopoly on this polymerase. The company can restrict production and control price.[13] For a university lab with finite resources, an expensive research material means that they can afford less of something else.

Intellectual property law can affect laboratory practice as well. The proposal to manufacture *Taq* in-house may be a dangerous move unless the Handelsman and Goodman labs have the capacity to fight Roche's legal challenge to the "experimental use exemption." In short, although university labs that oppose Roche's efforts to narrowly define the "experimental use exemption" and restrict the use of *Taq* might desire to confront the company's legal position, it would require time and money. Indeed, MacKenzie and his colleagues note, "Patent litigation is prohibitively expensive. For example, a firm can incur triple damages and costs for willful infringement" (1990, 70). Roche certainly has the legal expertise to defend its position, and the company is apparently willing and economically able to do so. Most university laboratories, by contrast, do not have lawyers on staff; a legal challenge to the Roche position, even if economically possible, would take scarce resources away from bench research. Under such circumstances, university laboratories are likely to acquiesce.[14]

The character of power here is rather straightforward. In his critique of pluralist theories of power, Lukes—following Dahl—defines the "one-dimensional view" of power: "A has the power over B to the extent that he can get B to do something that B would not otherwise do" (Dahl quoted in Lukes 1974, 11, 12). Cases of power of this variety may be rare, or politically or analytically less important than others. It is clear, however, that Roche's threat of legal action could prompt university labs *not* to do something they might do under different circumstances: that is, manufacture *Taq* themselves. Of course, if Roche is successful in the courts, the meaning of the experimental use exemption—the definition of who qualifies and under what circumstances—may become more limited in the law, making the kinds of considerations raised in the lab meeting unthinkable. In other words, if there is no tradition of using patented materials in university research without license, then such practice will not even be contemplated. If so, we move to a more complicated power dynamic in which a set of rules governing intellectual property practices is altered to the advantage of companies like LaRoche, and these rules are accepted as standard by all parties concerned.

Similar issues are raised by the case of Cre-*lox*P, a technology that can be used to manipulate the genes in mice for research purposes.[15] The technique was proposed in 1985 by scientists working for the chemical giant, DuPont. The tool utilizes the "natural gene splicing system" from a bacteriophage—a virus that infects bacteria—and adapts it for use in complex organisms. The technique "is based on two genetic elements of the P1 bacteriophage: a gene called *cre* that expresses an enzyme not normally seen in higher organisms, and a stretch of DNA called *lox*P. . . . When Cre encounters two *lox*P sites in a stretch of genetic code, it clips out the intervening DNA, along with the *lox*P sites, reattaching the ends to make a seamless strand" (Marshall 1997b, 24). This technology can be used to create "conditional mutants," mice in which a specific gene is bracketed for deletion in particular cells producing the Cre enzyme. It can also be used to create other genetically altered mice, including so-called straightforward "knockout" mice. These mice then become valuable research tools. They can enable scientists to better understand how particular genes work and what function they serve.

In 1990, DuPont was granted a patent on Cre-*lox*P to modify DNA in the cells of higher organisms (eukaryotes). DuPont is interested in regulating (or restricting, depending on whose perspective you listen to) the use of this powerful method for manipulating genes. Consequently, the company is insisting that scientists using Cre-*lox*P mice acknowledge DuPont's rights to the animals and commit themselves to sharing money made on discoveries in which the technology is utilized. Furthermore, scientists using the technology are restricted in their distribution of it to only those investigators whose institutions agree to the company's terms (Marshall 1997b, 24, 25).

In this tangled corporate web, the practices of university scientists might be affected in a number of ways. First, if payment of royalties is required for using the technology, expense may limit its use by cost conscious laboratories. Second, as then-director of the U.S. National Institutes of Health Harold Varmus suggests, these requirements constitute a significant burden for the individual scientist; they will slow things down. Finally, and also quite conceivably, what a scientist can do with any invention she develops using the Cre-*lox*P technology could be affected by DuPont's requirement that unspecified royalties be paid to the company on commercial discoveries facilitated by use of the tool.

These considerations can only be understood by recognizing the character of the relationship between DuPont and university researchers interested in using the Cre-*lox*P technology. It is a relationship of resource dependency. DuPont owns and controls a resource vital to the work of some scientists. The resource cannot be acquired on better terms from another supplier. Academic labs are, therefore, dependent on DuPont. Furthermore, academic scientists like workers in the Handelsman lab are unlikely to be in a position to challenge the restrictions imposed by DuPont. Drawing on the tradition reflected in the "experimental use exemption," scientists might object to DuPont's requirements. They are not likely, however, to have the economic wherewithal to seek legal redress for their grievance. As an alternative, they might use the technology without permission. In this case, they face the possibility of being sued. Again, like Roche, DuPont certainly has the capacity to enforce its position through litigation. Individual university scientists are unlikely to have this ability.

Another instance of property rights considerations potentially affecting the use of research tools arose in the late 1990s around a new form of contract called a "materials transfer agreement," or MTA. These are contracts that academic scientists must sometimes sign before they will be permitted to use research materials available to them from a colleague. In order to obtain and use materials subject to an MTA, scientists are sometimes required to accept restrictions on their freedom to publish, and the colleague interested in supplying a biological sample, reagent, or other material may be required to delay release until her own university has clarified property rights considerations.

An interesting case of the effect of an MTA was reported in *Science* magazine in the fall of 1997. According to the report, two University of California scientists asked a colleague at Oxford University in England to provide them with some mammalian DNA sequences that the Oxford colleague developed. The American scientists wanted the material for an experiment involving transgenic mice. Before Oxford would send out the genetic material, they asked the University of California researchers and their nonprofit sponsors to sign an agreement in which the scientists surrendered any property rights in inventions developed with the genetic material. In addition, Oxford requested the right to preview and comment on articles arising from the scientists' research. According to scientists interviewed for the article, these kinds of requirements are increasingly common in certain areas of the life sciences (Marshall 1997g, 212).[16]

Superficially, the case of the University of California scientists may appear quite different from the *Taq* and Cre-*lox*P cases that I have discussed. The first two cases involved corporate-university relations and formal agreements for use of technologies. The MTA case involves relations between universities. The *Taq* and Cre-*lox*P cases involve specific patented technologies. MTA cases might or might not involve proprietary material. But beyond such differences, these cases have a great deal in common. Presumably finding a different source of the genetic material they wanted would not have been possible for the California researchers. Consequently, their desire to use the Oxford tool in their research creates a relationship of resource dependence. The California researchers depend on Oxford to provide them with that material, and they are thereby constrained. The practice of their research is fundamentally affected by the asymmetry of this relationship. In addition, what will happen to the research after it is completed is also affected by this relationship. Unlike the other two cases, their research is subject to review. Unlike the *Taq* case, but like the Cre-*lox*P case, the California scientists' own intellectual property rights, and consequently, what they might do with an invention arising from their research is affected by their need to use this particular research material.[17]

Finally, these apparently disparate cases all reflect the increasing pervasiveness of a culture that demands intellectual property protection and promotes the narrowing of the intellectual commons. Indeed, the MTA episode shows perhaps better than the other two instances discussed that even when the relationship does not involve the economic interests of industry, universities are still affected by the commercial world. Universities take actions that reflect the commercial norms that increasingly shape academic decision making.

Norms in Science and the Ideology of Intellectual Property

In the next two sections of this chapter, I explicitly address the intellectual property practices of academic scientists, focusing on the Handelsman lab, in particular, and explore the ways in which the involvement of university scientists in the intellectual property regime can affect the daily working lives of these researchers. In this section, I traverse this broad topic by juxtaposing problems in the free flow of materials and information as they exist in relationships the Handelsman lab has had with others in the scholarly community with commonsense assumptions about the efficacy of patent protection and how these assumptions affect the behavior of university scientists.

My argument is that debate over the corrosive effect of UIRs on academic science has focused too tightly on what UIRs mean for the flow of scientific information in the academic community. As a consequence, administrators, scientists, and policy analysts have largely ignored the barriers to open communication that stem from other factors that have little to do with corporate involvement in the academic realm, and, indeed, very likely existed *before* the widespread emergence of UIRs in the biological sciences. At the same time, interested parties have typically overlooked how the pervasive culture of intellectual property—commonsense ideas about its efficacy—can affect the practices of scientists, independent of restrictions that might be directly and explicitly imposed within the context of a university-industry relation.

Over a half century after he first articulated it, Robert K. Merton's delineation of the norms of science stands as the clearest statement of the orientation that is supposed to mark science as a distinctive social institution. Central to the normative structure of science, according to Merton, is what he refers to as scientific "communism." For Merton, the term means the "common ownership of goods" (1973 [1942], 273). According to this norm, "The substantive findings of science are a product of social collaboration and are assigned to the community. They constitute a common heritage. . . . Property rights in science are whittled down to a bare minimum by rationale of the scientific ethic" (1973 [1942], 273). According to this position, secrecy is strictly forbidden, and Merton asserts that the capitalist patent system and private property in technology are incompatible with this aspect of the scientific ethos (1973 [1942], 275).

The logic of Merton's conceptualization has been sharply criticized, and a good deal of research challenges the empirical reality of the ethos of science.[18] Conceptually, for example, Mulkay (1980) shows that it is not reasonable to assume that a given norm has a single literal meaning. Instead, he suggests that interpretation of norms is context dependent, and thus, violation of the literal meaning of norms is likely in practice. Empirically, Mitroff (1974), for example, in his investigation of scientists doing moon-related research, points to the counter-norm of secrecy operating as a balance to the norm of communism.

Despite the wealth of critical research on the circulation and operation of these traditional norms, the earliest commentators on UIRs in biotechnology treated these new relationships as a virtually unprecedented threat to the norms of science. In his discussion of university-industry cooperation, for example, a former Yale University president, A. Bartlett Giamatti, wrote of the tension between the "academic imperative to seek knowledge objectively and to share it openly and freely" and the contradictory goal in industrial science to "treat knowledge as private property" (1982, 1279). Even those who recognize the questionable reality of these norms in the university context comment on the threat to scientific openness posed by the rapid expansion of the "university-industrial complex" (Kenney 1986, 108–111). What is more, according to one analyst, "it is fair to say that many scientists believe that these norms guide their practice" (Rabinow 1996, 13).

In my six months of participant observation in the Handelsman lab and a subsequent year of monitoring the lab, I found no evidence that the laboratory's relationship with industrial collaborators or the desire of lab personnel to "commodify" laboratory inventions led to secrecy or restrictions in the flow of laboratory materials to those outside the lab. In laboratory correspondence, Handelsman does keep WARF—the University's patent agent—apprised of the lab's intentions concerning presentation of laboratory research at scientific meetings or plans to submit publications. However, as one lab member said: "I think we are pretty good about not letting the patenting process dictate how we want to behave." Indeed, my observations confirmed this statement. In fact, on several occasions where related issues were raised at meetings with graduate students, Handelsman was clear that communication with other scientists was always the lab's priority. One graduate student said, sometimes "there is a little bit of a struggle . . . to get everything written up" for WARF before a scientific meeting, but another student said that he "was going to talk about" his latest scientific breakthrough whether or not intellectual property protection was assured.

Although the free flow of information has remained relatively unimpeded by the Handelsman lab's interest in patenting and commercial development of lab inventions, laboratory activities have on occasion been hampered by restrictions running in the other direction—that is, from outside the lab into it. In one instance, as I discussed in the previous chapter, a graduate student sent his bacterial strains to a company that performs analysis to determine a particular type of biochemical profile of bacterial strains. When he indicated to company staff members that he was unhappy with the work that had been done, a senior company official forbade him from using the results in published material. In another instance, a Ph.D. student contacted a scientist and requested a particular *E. coli* strain that the scientist referred to in a published paper. When the material arrived and the student found that it was not alive, he contacted the scientist requesting a fresh sample. He received no response. In still another case, one lab scientist told the story of a friend who could not get published strains from a competitor, "apparently because . . . [the friend] was doing research that could possibly undermine the competitor's conclusions." Finally, a student working on a lab project sought genetic material from a lab doing related work, but he never received it. "They haven't been too awfully helpful," the scientist volunteered, and he speculated that the overlap in their work could be one reason.

Stories of such difficulty are common sources of frustration and humor in the lab. In most instances, commercial motivations appear to have little to do with information and material flow restrictions. Sometimes the explanation may be inter-lab competition. In other cases, as one lab worker suggested, failure to maintain biological samples properly may make it impossible for scientists to respond to requests for samples. In still other cases, it may be simple laziness. Other reasons for barriers to the free flow of information reported in the science press include nationalism and the dissolution of collaborative undertakings (Mishkin 1995, 927; Cohen 1997, 1961). As I discussed in chapter 2, a 1997 study and a survey undertaken in 2000 suggest that many factors other than commercial ones affect the willingness of researchers to supply information and/or materials to other scientists (Marshall 1997, 525; Campbell et al. 2002).[19]

What is important is that although openness may be an ideal in science, it is often not fully realized. In the case of the Handelsman lab, it is impossible to know for certain what prompted other labs' slow or lack of responsiveness to requests for information and materials. But one can certainly imagine barriers—for example, the pressures of the "tenure clock," or a failure to maintain research materials properly—that predate, or are independent of, the introduction of UIRs to academic biology.

In the Handelsman lab, then, restrictions in the flow of information and research materials cannot be attributed to the lab's *direct* relations with science-based companies. On the other hand, laboratory practices, such as decisions about how to patent, what to patent, and why patenting makes sense, are shaped by lab members' acceptance or recognition of commonly helped ideas about intellectual property protection.

According to one graduate student, intellectual property issues come up "very seldom" in the lab, "Maybe once a month . . . , [these] issues come up in passing." Still, patenting inventions derived from lab research is a regular practice in the Handelsman lab. But this is not a money-hungry group, and individual gain is not prominent among the reasons that lab personnel cite to explain why they patent laboratory inventions. In both formal interview contexts and informal discussions, Ph.D. students, lab scientists, and Handelsman herself listed three primary reasons to pursue patents on lab technology. First, in times of shrinking federal research dollars, lab researchers see patenting as a potential source of revenue to support further lab research. One Ph.D. student noted that patenting is "good financially for the lab. . . . [T]he lab gets a certain percentage and then that can help fund other projects." Another researcher suggested that patenting "directs proceeds . . . back in the direction of where useful things came from." Second, lab workers believe that patenting may give the laboratory greater control over the development of their work into commercial products than they would have otherwise. One lab researcher said that if Handelsman patents an invention then "she's got sort of ultimate freedom to do with that discovery whatever she wants." Another noted that patenting "gives the creators of something more control over how it is developed." Finally, lab personnel are interested in their work having social utility, and Handelsman suggests, "If the invention is unlikely to be developed in the public sector, if it is not something that can just be used the way it is, but needs some major investment, then we tend to patent because it's the only way to ensure that a company or somebody will really take it and run with it commercially."

As to the first consideration, although providing financial support for continued research in the laboratory seems a worthy goal, the history of patenting at the University of Wisconsin makes realization of this goal seem unlikely.[20] Between its founding in 1925 and 1986, WARF received 2,400 disclosures from faculty. Of those, fewer than 20 percent were patented. Only ninety-five patents resulted in licenses, and only seventy-six produced net income. In short, only 3 percent of WARF patents between 1925 and 1986 produced income, and only 1 percent earned more than $100,000 (Blumenthal, et al. 1986, 1624).[21] As to the matter of control, as I discuss in the following section, evidence from correspondence between Handelsman and WARF suggests that after WARF is assigned a patent, researcher control over invention development and commercialization may be limited.[22]

If it is money and control they want, it may appear as if members of the Handelsman lab are wasting their time, but this judgment is hasty. It is the matter of social utility that makes it important for the lab to seek patent protection. To realize the market potential of their work, the lab must attract commercial interest. To do so, lab members believe it is necessary to patent their inventions and offer licenses to companies in a position to develop and market these possible products. Indeed, this is precisely how universities throughout the United States promote the commercial potential of biology-based inventions created by their faculty: they offer interested companies patent rights. Judging by the kinds of relationships companies seek with universities, firms are interested in exclusive intellectual property rights (see Etzkowitz and Peters 1991; Kenney 1985). In this context, lab members' own attitudes are irrelevant, since if company representatives believe in the importance of patent protection, they are unlikely to take an interest in unpatented inventions.

Although there is little evidence that intellectual property protection promotes innovation and widespread use of inventions (Dworkin, 1987), such a causal relationship is widely believed to be true (see Etzkowitz 1994, 392). Company representatives regularly speak of the importance of patent protection. According to Myriad's director of corporate affairs, Bill Hockett, the company would never have invested in the research that led to the discovery of the two genes associated with hereditary breast cancer if they had not been able to secure intellectual property protection.

According to Hockett, "Without the protection that the patent affords, a company could not invest hundreds of millions of dollars in getting it to the marketplace." He asserts further that "if anyone could use the results of our research and our investment, it would not be worth much for the people who invested in our company" (quoted in *The Scotsman* 1997). Etzkowitz and Peters suggest that small companies often do not believe they can "afford to take the risk of an unrestricted license, allowing others to follow up and easily copy them. They [believe they] need the protection of an exclusive license in order to take the risk of bringing a product to market" (1991, 163). Finally, without providing evidence, John Doll asserts that "without the incentive of patents, there would be less investment in DNA research, and scientists might not disclose their new DNA products to the public" (1998, 690).

Handelsman suggests that this need to patent may be "more important as an illusion than as a reality." But even if Handelsman and her colleagues do not believe patents work the way many in industry do, if she and her lab members wish to attract commercial interest in their work they must nevertheless operate as though the efficacy of intellectual property protection were true. Samuelson notes, "What matters is that most people in this society believe . . . [that the patent system promotes innovation], and that this faith guides people's actions" (1987, 12).

It is clear in this case that institutional context—here both a formal and informal, explicit and implicit set of intellectual property rules—shapes the practices of the Handelsman lab. From what she reads and hears, Handelsman has come to understand industry attitudes toward patent protection. She is interested in seeing her work commercialized, and she actively seeks financial support from industry for her lab's research. Thus, Handelsman uses patents to attract commercial interest in the lab's work. To do so, she must abide by informal corporate attitudes and implicit expectations. It also requires her to follow explicit formal rules for filing patent applications.

Unlike the previous cases of *Taq* and Roche, power is not explicitly exerted in this instance. Instead, a set of rules is widely taken for granted; they constitute a social common sense (Gramsci 1971), which shapes the practices of the Handelsman lab. It is beyond the capacity of individual labs to change these rules, and flouting them can prompt sanctions: university labs are likely to lose corporate patronage.

WARF-Warp: University Scientists and a University Patent Agent

Accepting the rules of the intellectual property regime does not permit one to acquire patent protection for one's inventions, and as Myers (1995) stresses, the process of applying for a patent and ultimately having patent claims accepted is time-consuming and expensive. Intellectual property law is a highly specialized and arcane aspect of the American legal system, and it would be nearly impossible for an unassisted scientist to successfully file for patent protection. WARF has the legal expertise and financial wherewithal to file patent applications on behalf of University of Wisconsin faculty. It is here that we see a final instance in which the structure of the intellectual property regime *indirectly* shapes the practices of the Handelsman lab.

WARF was established in 1925 as an independent nonprofit foundation to "administer patents and licenses resulting from research discoveries brought to it by University of Wisconsin faculty members and to use the income from such licenses to fund further research at the university" (Blumenthal, Epstein, and Maxwell 1986, 1621). After Professor Harry Steenbok's effort to obtain a patent on a process he developed for activating vitamin D in food by irradiating it, "friends of the university" proposed the establishment of WARF (Blumenthal, Epstein, and Maxwell 1986, 1622).

To enter WARF's world one must enter on the organization's terms. WARF will not pursue just any faculty disclosure, but only those that Foundation staff believe will be successful and profitable. WARF is offered approximately two hundred faculty disclosures per year. Of these, WARF converts fewer than half into patent applications. Ultimately, barring independent resources[23] or the interests of a company willing to pursue patent protection in collaboration with a university scientist, Wisconsin researchers are limited in their pursuit of intellectual property protection and associated licensing to only those inventions that are of interest to WARF.[24]

This can be a problem. According to Handelsman, WARF is "very well equipped to cut certain kinds of deals with companies." But Handelsman finds it unfortunate that the Foundation has "extremely good experience with only a few kinds of technology." And she is "not convinced that they know how to recognize really good technology that doesn't look like those" with which they already have experience. She is "not convinced that they recognize the big hits, the real innovations." Given this situation, one of Handelsman's collaborators said that the lab considered pursuing intellectual property protection independently, but rapidly realized that this path was not "realistic." And correspondence between WARF and Handelsman indicates that the relationship between the Handelsman lab and the organization has sometimes been strained.

Although WARF has been reasonably accommodating, documents make it clear that decisions concerning invention licensing are up to WARF. In one instance, WARF agreed not to enforce patent protection on a lab invention if and when the invention was used in a restricted way by Wisconsin farmers with whom the Handelsman lab hoped to collaborate.[25] According to a letter from a member of WARF's staff to Handelsman,

> WARF will forbear from enforcing any patent it may obtain from this technology against individual Wisconsin farmers practicing the technology with respect to biocontrol agents for use on their own farms. WARF will similarly forbear such enforcement against farmers' cooperatives and other small businesses serving Wisconsin agriculture if they are located in Wisconsin and if their practice of the technology is limited to alfalfa and to the county in which their principal place of business is located and any contiguous county. (March 20, 1992)

The Foundation made clear, however, that "WARF's keeping itself free to permit in-state alfalfa practice of the technology is not the same as guaranteeing to affirmatively *allow* such practice" (March 20, 1992).

In other correspondence, WARF asserts that a company licensing a Handelsman lab invention need not provide the lab with information concerning the use to which the invention will be put. According to a letter from a WARF licensing associate to Handelsman, the company was "not willing to disclose what their particular needs were, only that it was outside the agricultural field of use" (November 27, 1991). Of course, this means that the lab loses control of its invention. Finally, the breadth of the patent drawn by WARF and the nature of the licensing agreement into which the organization enters with a commercial concern may restrict the freedom of movement Handelsman has in seeking further corporate support, as it appears to have done in one case. As a WARF staff member acknowledged to Handelsman: "The exclusive rights to which . . . [the company] has an option clearly were logical to grant at the time they were granted. Just as clearly, those rights make for present problems in your attempts to secure funding for additional research projects where patents on inventions that might be made in that additional research are likely to be blocked by the existing patents and applications covering your past research" (November 21, 1991).

If the patent claims are broad and a particular patent is licensed to a company, Handelsman may be prohibited from offering future related inventions to another company in her search for additional lab support. At a time when government support for research does not match demand, this kind of limitation could be significant for a lab like Handelsman's, which had a budget of $250,000 to 300,000 annually in the mid-1990s (see Etzkowitz and Peters 1991, 149). As I have noted, although the lab would receive compensation if its licensed inventions were profitable, most WARF-patented inventions during roughly the organization's first seventy years of history have not been big moneymakers (Blumenthal, Epstein, and Maxwell 1986).

For a university laboratory at the boundary between basic and applied work and at a time when research budgets are uncertain, patent and licensing activity may *indirectly* underpin work at the laboratory bench and influence research both financially and practically. If the sources from which a lab leader may seek funds, and what she can offer potential corporate collaborators or patrons are limited, this may affect the kind of research she and her staff can do in the lab. If a university patent agent decides lab collaboration with farmers—or anyone else for that matter—is not in the interest of the university, the agent may be in the position to restrict this research cooperation.

The University of Wisconsin is unique in permitting its faculty to seek patent protection on their own. Of course, cost would prevent a typical university scientist from pursuing intellectual property protection independently, but throughout the United States this is not an option, in any case. As university employees, scientists at universities other than Wisconsin must work with their own university patent agents. Considering the growing importance of intellectual property protection in the biological sciences, university researchers will increasingly be pressured to cooperate in the patenting of their work. Once they have accepted the need, or acceded to the pressure, to pursue intellectual property protection, university scientists' role in determining how to pursue this protection and with what limitations will be constrained by the their inability to seek patent protection independently. This is true despite the expectation that institutions like WARF should serve as a buffer between universities and industrial firms (Etzkowitz and Peters 1991, 157). Here we have a case of power of the most basic variety: the resources (money and legal expertise) at the disposal of WARF and similar patent agents can lead university labs to act in ways they might not if rules permitted university scientists to pursue patents on their own and they had the resources to do so.

Conclusions

Hardly a week goes by in the science media without a story about a new patent, a new lawsuit, or a new intellectual property policy. Scientists, analysts, and policymakers are quoted bemoaning the restrictions on the flow of information that result from the commercialization of the biological sciences. The question is: how can we best understand trends in intellectual property rights? I suggest that it is useful to understand the U.S. intellectual property regime as a system of rules and resources that empower some actors more than others. In this system, to a significant degree, commercial research tools suppliers are more powerful than university scientists. They can constrain the research practices of these academic researchers by working to further limit the experimental use exemption to patent law and by threatening legal action against scholars who challenge restrictions on the use of biological research tools.

University patent agents are also typically more powerful

than academic researchers. They will represent university scientists, but the terms on which they do so are not always to the best advantage of the scientists. Indeed, the responsibility of organizations like WARF is not to serve the interests of individual labs or scientists, but to improve the economic future of their university. This objective may frequently be at odds with the interests of individual university labs and scientists. Finally and importantly, whatever academic scientists' opinion of intellectual property protection, common beliefs about the virtue of patenting, whether true or not, make efforts to skirt the intellectual property regime in any attempt to promote the utility of the products of scholarly research beyond the walls of the university unlikely to succeed. Put simply, because of widespread belief in the efficacy and necessity of patent protection, the interest by companies in university inventions will typically depend on patenting and licensing.

My approach in analyzing intellectual property protection in academic biology differs from the approach typical in policy analysis of UIRs, and also differs from work central to theoretical debates in science studies. While I do not dismiss the importance of the effects of *direct* and *formal* university-industry partnerships in restricting the flow of information, there are many noncommercial factors that can inhibit this information stream. Just as importantly, the *indirect* and *pervasive* influences of the character of the intellectual property regime on academic biology are crucially important to understanding the current character of academic biology. As to debates in science studies, while I happily acknowledge the importance of discourse in understanding social phenomena, we must understand the historical and institutional bases that give discourse or those who draw on its power. Academic scientists can say whatever they want about patent protection, but if social common sense suggests that without patenting there will be no innovation or commercialization, then there will not be. Finally, while it cannot be doubted that developments in the realm of intellectual property protection do construct or alter social worlds, if we focus only on these constitutive processes, we overlook how already constructed and deeply entrenched structures—such as the intellectual property regime—shape the daily practices of the actors.

6

It Takes More than a Laboratory to Raise the World

In the early 1990s, an interesting and provocative hypothesis emerged from the Handelsman laboratory. The idea was that bacterial strains with biological control activity would be more effective at suppressing plant disease at the site from which they were isolated—typically a farm field—than would strains imported from another site. That is, strains adapted to a particular site would be more powerful biocontrol agents than strains from elsewhere. This hypothesis led to discussions between members of the Handelsman lab and social scientists interested in rural development issues; it ultimately sparked a year long collaboration. After a year, the collaboration was terminated.

In the mid-1990s, following a series of "successful" interdisciplinary seminars on pathogen and disease control at the University of Wisconsin and discussions between Handelsman, the dean of the College of Agricultural and Life Sciences, and others, a decision was made to initiate an Institute for Pest and Pathogen Management (IPPM). This would be an interdisciplinary program that would support graduate training and basic and applied research. Funding would come from a range of sources, but industry would be a major supporter of this program.

In this chapter, I use these two cases as the basis for a discussion of the relationship between science and the social world—the connection between the technical and the social. I continue to pursue the argument from the previous chapters that it is impossible to understand the dynamics and character of university biology today without understanding the social environment in which the university and university science are embedded. My claim has been that insofar as the Handelsman lab is representative, the practices of university sciences are shaped by the world in which they are situated, and that commerce, in the broadest possible sense, is a significant feature of that world. I add to this argument a claim that runs counter to a position commonly held among researchers in the science studies field. Many in the area contend that it is impossible to separate the social and the technical, and more difficult still to give one analytical priority over the other.[1] I suggest that to the contrary, although the social and technical are integrally related, for analytical purposes it is often profitable to make these distinctions and to examine the hierarchical priority of the social.

In this chapter, first, I outline and critique a paper by Michel Callon, which is at once a manifesto for science and technology studies scholars who contend that the social and the technical cannot be usefully disaggregated and is an example of the kinds of arguments posited by researchers who hold this position. Next, I describe the Handelsman lab's bioregional biocontrol project and point to the social factors that hindered its successful development. Third, I profile the University of Wisconsin's Institute for Pest and Pathogen Management and draw attention to the social-organizational factors that made that initiative promising. Finally, I consider some of the implications of my comparison of the bioregional biocontrol project and the IPPM effort for understanding the factors that shape technoscience.

Callon's Car Crash

In an often-cited article, Michel Callon (1987) tells the story of efforts to develop and commercialize an electric car in France. He uses this case to question "the claim that it is possible to distinguish during the process of innovation phases or activities that are distinctly technical or scientific from others that are guided by an economic or commercial logic" (1987, 83). According to Callon, "This distinction is *never* . . . clear-cut" (1987, 84; emphasis added). He suggests that a successful electric car demands "electrons that jump effortlessly between electrodes; . . . consumers who reject the symbol of the motor car and who are ready to invest in public transport; the Ministry of the Quality of Life, which imposes regulations about the level of acceptable noise pollution; Renault, which accepts that it will be turned into a manufacturer of car bodies; lead accumulators, whose performance has been improved; post-industrial society, which is on its way" (1987, 86). Callon concludes, "None of these ingredients can be placed in a hierarchy or distinguished according to its nature." The activist in favor of public transport, asserts Callon, "*is just as important as* a lead accumulator, which can be recharged several hundred times" (1987: 86; emphasis added). In short, Callon contends that it is not possible to meaningfully distinguish between social and technical factors in an effort to understand the development of new technologies. Neither area can be granted analytical or explanatory priority.

I suggest that this popular analytical framework is a barrier to understanding the current character of university science and technology. Of course, few would deny that a successful technology must *simultaneously* work in "technical" terms and develop in a supportive social environment. However, this is not the same thing as asserting that we cannot distinguish the social and technical. Callon never examines what it means for something to be "distinctly" technical or social. He does not entertain the possibility that while boundaries between the social and the technical may, at times, be indistinct, there might be good analytical reasons for highlighting the differences. I contend that some factors play a more important role in defining the contours of our social fabric than others, and, indeed, some are more important than others in determining the course of socio-technical development.

Callon points to the "technical" failure of the automobile battery as crucial to understanding the demise of the French electric car and views this as a factor that social scientists have traditionally overlooked. In the case of bioregional biocontrol, it is, of course, necessary that scientists be able to determine subtle regional differences in the efficacy of biocontrol agents. In short, the "technical," so often ignored by social scientists, is clearly important. However, at a most basic level, scientists need financial support to determine whether the hypothesis is tenable. Without that support, the workability of bioregional biocontrol is irrelevant. Thus, in such instances, economic resources have temporal and consequently analytical priority over "technical" feasibility. If funding or its absence can be linked to the broad character of social organization or societal values, then this too has explanatory priority; thus, it is clear that the social and technical can be separated for analytical and explanatory purposes.

There is a second aspect of Callon's argument that can be scrutinized with the information from the bioregional biocontrol project in mind. As part of his critique of sociological approaches that ignore the technical and/or arbitrarily separate it from the social, Callon suggests we can learn something from what he describes as the engineer-sociologists in his case. He suggests that the engineers from Electricité de France (EDF) "presented a plan for the VEL [an electric car] that *determined* not only the precise characteristics of the vehicle it wished to promote but also the social universe in which the vehicle would function" (1987, 84; emphasis added). Callon asserts further that "by predicting the disappearance of the internal combustion engine as a result of the rise of electrochemical generators and by ignoring traditional consumers so as to better satisfy users who had new demands EDF not only *defined a social and technological history* but also identified manufacturers that would be responsible for the construction of the new VEL" (1987: 85, emphasis added).

These formulations are odd. At one level, the claim that these engineers are attentive to both the social and the technical and that they properly regard them as intertwined is trivial. As I discuss below, few people would be unaware that a successful technology depends not only on the technical workings of the invention, but also on a hospitable economic situation. At another level, Callon seems to imply that these engineers have a kind of efficacy—a capacity to construct the world—that seems improbable. It seems more plausible to me to suggest that Callon's engineers made assumptions about the environment in which their technology would succeed than to contend that these engineers were capable of "determining" or shaping the social universe. Similarly, I am happy to grant that Callon's engineers made assumptions about how the world would need to be reorganized in order for their electric car to succeed. But that they, independent of unstable social conditions and a powerful social movement, could fundamentally shape the social environment within which technical development takes place seems highly implausible.

What is more, Callon does not even provide evidence that his engineers explicitly specified what they thought the world should look like, or that they had a conscious and active strategy for shaping events.

As I have argued throughout this book, researchers in the Handelsman lab work in a world that is not of their making. They cannot easily alter aspects of their environment to suit their needs, but must accommodate themselves to what it is. In this context, the bioregional biocontrol case provides a portrait of a very different variety of inventor-sociologist than Callon's. The scientists in the Handelsman lab recognize that they operate in a world of constraints, and they often allude to the difficulties they confront in their efforts to circumvent these barriers.

Socio-reason / Bioregion

A graduate student from Handelsman's lab described the genesis of the bio-regional-biocontrol idea:

> I was working the with bacteria phage P7, and . . . we thought there was a possibility that the phage might be carrying the genes for zwittermicin A production on it and so I was trying to study that. To do that, I had to find other strains that I could infect with the phage. . . . What I wanted to show was that they weren't producing the antibiotic and then when I integrated the phage they would and what I found was that all the strains that I could infect with the phage already were producing the antibiotic. That was the first time that we really found that any strain other than UW85 produced the antibiotic, and at the same time, all of the sudden we immediately had a tool for finding them because all we had to do was to look for strains that were sensitive to this phage. . . . That was really a surprise.
>
> It hit me pretty immediately that that was pretty neat, and I remember going into Jo's office and just spouting off about how neat this was and how we could go to . . . different soils and maybe we could find strains in those soils that were like UW85, but would be adapted to those soils.

Informal discussions between Handelsman and a sociologist affiliated with the Agricultural Technology and Family Farm

Institute (ATFFI)[2] at the University of Wisconsin led to a year-long collaboration. The biologists and social scientists attempted to match ideas of appropriate bioregion—the size at which the advantages of a locally isolated bacterial strain would diminish (that is, answering the question: How large is "local"?)—with the most appropriate form of production and distribution system for the technology. The collaboration was made possible by unencumbered funds available through the ATFFI, and initial discussions led workers to pin their hopes on a decentralized system of production.

Several biological and socio-economic hypotheses along with substantial elaboration and justification for the bioregional biocontrol project are provided in project documents produced by the research collaborators. As justification for a bioregional biocontrol approach, the investigators suggested that

> despite some striking successes in biocontrol of crop disease, the field in general has been disappointing and results have been variable, offering little in terms of reliable disease management. A partial explanation for the variability may lie in the general expectation that one strain of a biocontrol agent should perform well in diverse conditions where there are extreme variations in soil and climate conditions or in the complexity of pathogen strains and host genotypes present. ("An Approach. . . ." n.d., 2)

As to the priorities of the project, one document suggests, "Perhaps the most critical aspect of the project is to test the 'regional specificity' of the biocontrol [agent]—that is, in how wide or narrow a geographic area are the bacteria effective" (Stevenson et al. n.d., 2). But if this suggests the authors of this proposal are giving some sort of priority to technical matters—to questions of biology—this claim is followed by a discussion that points to the tight links between matters of production organization and problems of biology:

> If the most effective biocontrol agents are those that are isolated from the fields in which they are to be used, then a customized service might be needed to isolate biocontrol agents from farmers' fields, formulate them, and return them as a product to the farmer. The socioeconomic corollary to a bioregional approach to disease biocontrol

is a decentralized product development and distribution enterprise. (Stevenson et al. n.d., 2)

The link between the character of a bioregion and the appropriate form of production and distribution organization is elaborated by the collaborators in other documents. They outline five possible models. These include: a system of on-farm production and use; a small-scale, local proprietary business or cooperative that would produce the biocontrol agent for farmers; a large-regional cooperative system or proprietary business; a centralized state-run service organization; or some hybrid combination of these (Hendrickson et al. n.d., 1).

During the year-long collaboration, a limited amount of work was done. The social scientists assembled focus groups to learn the attitudes of agricultural sector workers who might be involved in testing a new product and new organizational form. They received support from an agricultural cooperative near Green Bay, Wisconsin. The co-op expressed an interest in testing a biocontrol product on farms and a willingness to market the product utilizing a hybrid development model. The biologists undertook a limited test of bioregion size across selected sites in Wisconsin.

After the ATFFI money dried up, the project ended with no conclusive research results and without a single model production system up and running. Project participants proposed related but distinctly different explanations for the demise of the effort. One social scientist put the problem directly in terms of the technology. He said, "The two big questions that it seemed to us we were waiting on were the questions about does it work, and if it does work, what's the boundary, what's the unit here, what's the region?" This collaborator said that when he asked Handelsman about the nature of the bioregion, she told him that the data that they had amassed would not permit them to provide any definitive answer to questions about the nature of a bioregion. The data, as the researcher recalled, did not show significant differences in efficacy between isolates native to one site and those from other sites. At this point, according to this investigator, he remembered Handelsman saying "We're having trouble keeping this thing funded."

The subtly different answer of a biologist who worked on the project in its early stages is telling. According to the scientist, two things worked against bioregional biocontrol:

> First, in my opinion the best way to test this is to go to radically different bioregions, Honduras, Louisiana, the Netherlands, Wisconsin—something like that. To do this right you'd need $ and really reliable, dedicated collaborators. It never happened. Looking at five sites in Wisconsin is something we could handle [with the available resources], but we were less likely to find dramatic results, and we didn't get dramatic results. Second, people that have been interested in our strains, outside companies that are testing them in the field themselves, are looking for Mr. Goodstrain. At best, they might let the idea of bioregional biocontrol influence their choice of strain sources, but they aren't interested in testing the hypothesis. They want a strain that gives them a big yield increase. (correspondence, December 2, 1997)

The researcher pointed out that to test the bioregional biocontrol hypothesis is not a trivial matter. He suggested that the approach is not likely to point to extremely sharp differences. At best, he said, "you are looking for a *tendency* for strains to be better. You certainly need to test multiple strains over multiple years at multiple sites and pray for disease pressure." The scientist concluded by pointing to the pivotal role of funding in undertaking an adequate test of the bioregional biocontrol hypothesis. He noted that to properly test the hypothesis would require much more money than would be needed to find a few generally effective biocontrol agents.

Handelsman herself has ideas about why the project died. First, she said, "Although the sociologists talked a lot about doing technology assessment before or during biological research to try to shape outcomes, no one was really willing to do it." Furthermore, the economists involved in the project, according to Handelsman, thought in terms of traditional models of economics and consequently pushed scientists to think in terms of "the largest [biologically feasible] bioregion to maximize economies of scale." This, Handelsman said, "puts us right back in the 'Monsanto model.'"[3]

Like the central researcher in Handelsman's lab working on the project, Jo Handelsman's analysis of the failure of the project puts emphasis on the problem of adequately defining a bioregion. Like that scientist, she suggests such definition is a complicated matter: "You would need to test a group of strains in plots over increasing radius size from the site at which the strains were isolated, and try to correlate performance with distance from point of origin. Then you'd have to replicate this over and over. It would require hundreds of research plots" (correspondence, December 3, 1997). Such an effort would, according to Handelsman, cost "literally millions of dollars," and she and her colleagues had had trouble finding support for a small-scale project. Thus, the unwillingness of funders to support the project and the unwillingness of the social scientists to proceed without a firm definition of bioregion led her to give up on the project.[4]

Interestingly, despite the complexity of defining a bioregion, peer evaluations for two of the grant applications submitted by Handelsman and her social science collaborators do not point to this as a problem. One reviewer said the "plant pathology components of the proposal" were "technically strong." Another reviewer characterized the "quality of the science" as "excellent." Still another described "the plant-microbe component of this proposal . . . [as] very strong" but said the economics component was very weak. This reviewer recognized a difficulty the social scientists noted themselves: it is hard to "develop an economic model without clearly knowing more details of the product and/or services which are to be provided and the economic incentive for adoption of the 'product.'"

In querying a preproposal, the director of a rural development organization looked at the economic problem not in terms of the group's failure to specify compelling models because they lacked data on the nature of the bioregion, but instead questioned the economic feasibility of the bioregional approach. What, he said, "is the basis for the assumption that locals will pick up on and commercialize the regional biocontrol of agricultural pests?" He thought it plausible that the approach might be taken up by larger agribusiness. If this were the case, then a regional economic development organization would not be interested in the project.

How then should we assess the failure of the bioregional biocontrol project? If we were to assume that the peer review process produces meritorious outcomes, then we would have to conclude that compared to other projects with which the Handelsman proposal was competing, theirs was just not that good. But this conclusion is too easy.[5] On the one hand, there is simply no objective way to assess the project's intrinsic merits. Indeed, its merits or weaknesses cannot be understood intrinsically but must be understood in relationship to the world in which the Handelsman lab is embedded. I have argued that the context within which the Handelsman lab operates constrains the practices of the lab in a range of ways. Here, it is possible to delineate several ways the social environment in which the lab works shaped the failure of this project. First and most obviously, research depends on money. In thinking about the relationship between research and funding and contrary to the position of many in science studies, including Michel Callon, the social (or more precisely the economic) can be understood to exist *temporally* prior to the technical. Without funding, there is no research, and so there is no technical matter. This, of course, begs the obvious question: why was no support for the project forthcoming? On the one hand, we might say that funders were unwilling to provide support without data, thus suggesting the priority of the technical, but there could be no data without support. Callon might point to the inextricable bonding of the social (funding) and the technical (data). On the other hand, however, given Handelsman's excellent funding record over the years—recall that she managed to support a lab with annual costs of over a quarter of a million dollars in the mid-1990s—we should perhaps look at the reluctance to support the research as itself a constraint. Here, we might view the project as innovative but risky. Assessing the plausibility of the bioregional biocontrol hypothesis would require, as both the lead biologist on the project and Handelsman point out, extensive testing and consequently lots of money with no promise of payoff. Here, the structure—the norms and history—of the research funding economy is crucial. Research patrons tend to be risk averse (see Chubin and Hackett 1990). They are more likely to fund sure things than high-risk projects, even from established investigators. One historian notes that the task of assessing proposed research is especially difficult when researchers propose studies where "the probability of the fruitfulness of [the] research [is] of a type without reliable precedents" (Abir-Am 1988, 153). Again, this points to the priority of the social.

Beyond issues of norms of research funding, interdisciplinary research faces a particular set of difficulties. Among these is the resistance of peer reviewers to original interdisciplinary research. According to one historian, peer reviewers "with conventional conceptions of the rightness of research within disciplinary boundaries, from institutions in which these assessments were established, and from institutions which . . . [are] sensitive to criticisms from those who are affronted by unconventional scientific plans" are likely to be skeptical of interdisciplinary work (Abir-Am 1988, 176). Although much lip service is given to the importance of interdisciplinary research in efforts to solve complex problems, interdisciplinary investigations face entrenched and powerful disciplinary resistance (see Campbell 1969; Saxonberg and Newell 1983; Bechtel 1986). Historically, interdisciplinary programs have faced funding difficulties (Abir-Am 1988), and have been unstable or short-lived (see Solovey 1990). In the case of the bioregional biocontrol initiative, investigators were proposing to wed aspects of the biological sciences and the social sciences. The gap between these is probably larger than in most interdisciplinary initiatives (Whitley 1984; Fuchs 1992).

In the differing interpretations of the project's failure alone we can see the strain among collaborators created by different perspectives on the project. Indeed, in a status report produced by the major collaborators on the project, they themselves recognize the impediments to interdisciplinary research of this type: "The multidisciplinary teams that will drive technology development in new directions are difficult to develop and maintain. The clash of cultures of the social and biological scientists often prevents intellectual progress. The working styles, priorities, and incentives of the two groups can present substantial barriers to collaboration" (Stevenson et al. Status report, n.d., 9). In all of these factors, it is the social—the nature of research funding and the organization of intellectual work—that explains the failure of this project.

But let us assume for a moment that the project had been fully funded and that researchers had been able to define clearly and compellingly the boundaries of the bioregion. Let us imagine that disease suppression was best when bacterial isolates used to create biocontrol agents were taken and applied to the same farm, or at most, to the same county from which they were originally isolated. Let us suppose that biocontrol agents isolated at the farm or county level were substantially better than those imported from nearby regions or states. In this case, the collaborators on this project would have faced social constraints of a very different variety.

Recall that the "socioeconomic corollary" of the bioregional biocontrol hypothesis is a decentralized product development and distribution system. We must understand, as the collaborators in this project did, that the likely success of such a system would be constrained by the particular economic environment in which it was undertaken. Even with the emergence of "lean" and "flexible" production, large centralized enterprises dominate the U.S. economy (Harrison 1994).[6] Indeed, this has not changed much over the past several decades. This is as true in agriculture as in any other economic sector. In farming, economic concentration has meant fewer, larger farms. The pace of concentration has accelerated in the post-World War II period. While the number of farms in the United States declined by only 8 percent in the period from 1915 to 1945, it decreased by 55 percent—from 5.9 million to 2.8 million—between 1945 and 1975. Concentration is also occurring in the agrichemical industry.[7] In the mid-1970s, there were some thirty U.S. companies developing pesticides. This number was down to twelve by the late 1980s and is expected to drop to six in the coming years. There is also the trend toward vertical and horizontal integration. The top pesticide producers are also important agricultural seed companies (Busch et al. 1991, 23, 24). Farming, farm-input, and food-processing industries all follow a mass-production model. When they contemplate investment in bio-pesticides, their comparison is "fast" and "cost-effective" synthetics that are mass produced and not adjusted to suit local variation (Groves 1998, D7). Indeed, variability in efficacy—perhaps associated with variability in bio-region—has generated skepticism concerning biocontrol agents among members of the agri-chemical industry.[8]

In the various documents produced as part of this project, the social scientists and biological scientists who worked on the effort refer to an "appropriate" system of production depending on what the biological research shows about the nature of bioregions.

"Feasibility" is also tied to the character of bioregions (Hendrickson et al. n.d., 2). But as the researchers indicate that they are clearly aware, the type of production and distribution system that will succeed will be shaped in substantial measure by the economic context in which such organizations would operate. First, there is the issue noted in chapter 2 that "biocontrol will need to offer farmers decided advantages [over chemical control] to be competitive" (Stevenson et al. n.d, 7). Thus, success of this technology will be shaped by the history of chemical industry dominance of agricultural pest control. But more important for this specific effort, the researchers recognize the economic structure of the agricultural sector. The collaborators note in an unpublished status report, "The large-scale, proprietary business model (being the traditional avenue for technology development) functions as the 'standard' in this project against which the costs and benefits of other models are compared" (Stevenson et al. n.d., 5). In this report, the involved researchers add, "A biocontrol technology that is simple and inexpensive may be very attractive to farmers but not provide sufficient profit potential for rural businesses" (Stevenson et al. n.d., 7). In fact, this is a complicated issue. Decentralized technology does not neatly fit the centralized mass production model that dominates the agricultural sector. Moreover, the substantial resources of large corporations could permit an agrichemical firm competing with decentralized biocontrol producers to operate at a loss in order to undersell its small competitors.[9] Such enterprises would be better equipped to promote their product through both formal and informal means. Formal advertising campaigns would likely be beyond the economic means of individual decentralized producers; coordinated campaigns would face organizational obstacles. The networks of relationships between farmers, agrichemical distributors, agricultural extension agents, and university researchers could conceivably make it difficult for the decentralized system to break through established bonds of loyalty.[10]

These social structural barriers to success would be substantial, but there are even more difficulties for a project of this type. As I discussed in the previous chapter, matters of intellectual property protection are significant facts of life in American universities today. Recall that the Wisconsin Alumni Research Foundation (WARF) was not entirely happy with Handelsman's proposal for the bioregional biocontrol project when they learned it involved circumventing the patent they had succeeded in securing for the lab-developed technology for isolating biocontrol agents from soil. When I asked the primary biologist working on the bioregional biocontrol initiative about the demise of the project, he was quick to point out that "Despite all the WARF brouhaha, they gave us the green light to test whatever we wanted to, so I don't think we can really blame them" (correspondence, December 2, 1997). And, indeed, in a letter to Handelsman, which I quoted in chapter 5 (correspondence, March 20, 1992), a Foundation staff member reported the organization's decision to refrain from enforcing the relevant patents as long as the "technology with respect to biocontrol" was used only under very limited conditions in Wisconsin.

But this letter does not exactly give the lab and its collaborators a free hand. First, it is presumably the case that if lab workers found certain *Bacillus cereus* strains effective biocontrol agents against diseases that affect crops other than alfalfa they would want to pursue these. Second, allowing use in Wisconsin alone would not be adequate to enable full development of the bioregional biocontrol concept. There would need to be test sites well beyond the boundaries of Wisconsin or even the Midwest. In addition model production systems would need to be developed outside the state of Wisconsin. Finally, WARF's willingness to "forbear" enforcement in a limited area is not, as the organization notes, the same as "guaranteeing to affirmatively *allow* such practice" (correspondence, March 20, 1992). WARF's interests are by no means identical with the Handelsman lab's and those of its social science collaborators. The Foundation aims to maximize royalties and provide research support to the University of Wisconsin. If this technology turned out to be highly profitable within the existing system of production and distribution, WARF would be unlikely to allow workers on the bioregional biocontrol project to freely and fully develop their system.

In the project status report, Handelsman and her colleagues recognize the possibility that developing the bioregional biocontrol technology might present intellectual property problems. In this document, the researchers note that "WARF's responsibility is to generate income." And they ask, "can WARF be expected to license technologies in a manner that may benefit farmers or small rural businesses *rather* than maximize royalties[?]" (Stevenson et al. n.d., 4). In a research proposal, the collaborators suggest, "We will attempt to assess the compatibility for WARF's charter with the development of low-profit products for agriculture. The outcome of that assessment will potentially drive consideration of new models for intellectual property rights for university discoveries intended to benefit farmers and rural communities in Wisconsin" (Handelsman et al. n.d., 1). In this document, the writers suggest the possibility of assigning patent rights to a party other than WARF, and of not patenting new technology resulting from this project. In either case, the collaborators would confront the constraints imposed by the intellectual property regime. On the one hand, there is the cost of pursuing patent protection. As I have noted, in the mid-1990s WARF expended $17,000 in pursuit of each patent, and this is not a trivial amount of money for a university research lab and a group of social scientists. Similarly, it seems a cost that small developing businesses would have trouble bearing. Thus, if the workers on this project were to conclude it was necessary to patent their new technology, resource considerations might very well force them to work with WARF, and in doing so they would be required to pursue intellectual property protection on WARF's terms. On the other hand, if they were to pursue an alternative path and seek to avoid patent protection, they might very well confront the accepted common knowledge concerning intellectual property, which I discussed in chapter 5. In short, if business people believe that patent protection is necessary in order to profit from the development of an invention, they will be unlikely to pursue development of an unpatented technology.

Beyond illustrating the social constraints that face university scientists who seek to produce useful technologies and providing evidence that contradicts Callon's claim that it is impossible and unproductive to separate the social from the technical, this case provides a portrait of inventor-sociologists that differs greatly from the picture Callon sketches. It is one with quite different implications. What makes Callon's engineer-sociologists superior to traditional sociologists is their understanding of the world: it is a world that contains both entities sociologists normally consider (for example, consumers, social movements, and government agencies) and accumulators, fuel cells, electrodes, and the like: those things that sociologists normally ignore. But as I noted above, Callon is wrong to suggest that his engineers "determined" the characteristics of the social universe in which their vehicle would function, or that they "defined a social and technological history" (1987, 84, 85). Instead, the engineer's Callon described *made assumptions* about the world in which their vehicle would succeed. Unlike that of the collaborators in the bioregional biocontrol project, the analysis by Callon's engineers does not appear to have been explicit and self-conscious. In her assessment of the economists with whom she worked, Jo Handelsman showed an awareness of how the production and distribution model implicit in the bioregional biocontrol hypothesis contrasts with the traditional "Monsanto model." The lead investigator on the bioregional biocontrol project similarly indicated an awareness of how this project constituted a challenge to existing economic organization. In discussing the difficulty the lab had with WARF and one of its corporate collaborators when contemplating this project, a scientist working on the project spoke jokingly of a decentralized production and distribution system: "You know—that's communist. That's not the great capitalist system."

Lab workers and their collaborators had a highly analytical perspective on the environment in which their project was being undertaken, and this is amply illustrated in project documents. The collaborators show an awareness of how an economic sector dominated by large corporations could hamper the success of their project. They show an understanding, furthermore, of how the project could be shaped by agrichemical industry domination of pest control in the postwar period. They also recognize the barriers to their project's success posed by the intellectual property regime. Finally, unlike Callon or his engineer-sociologists, they are able to separate the social and the technical, and to understand the priority of the social—in the form of resources and the orientations of grant reviewers—in determining their ability to assess the technical feasibility of their hypothesis.

Instituting Pest and Pathogen Management

Several months after I ended my full-time stay at the laboratory, Handelsman told me about a new project in which she was involved. She was working with the dean of her college and others to create an institute for pest and pathogen management. A brochure published a year after our discussion, in March of 1997, describes the Institute as a mechanism for

> *building bridges* by bringing together a critical mass of faculty and students at the interfaces of disciplines that pertain to pest and pathogen management. It will foster collaboration between basic scientists attempting to delineate the intricacies of biological mechanisms and scientists who apply basic knowledge to real world problems. The Institute will also foster interdisciplinary research among public and private sector collaborators, and train students to conduct research at the interfaces between disciplines. (College of Agricultural and Life Sciences 1997, 2)

The Institute, according to this document, will support research in agriculture and medicine. It lists five "initial project areas": the molecular and ecological components of resistance management, the chemical ecology of biological control, the implications of pest and pathogen management for food safety, information technologies for integrated crop pest management, and farm system diversity and pest pathogen management.

The brochure highlights the importance of the program's efforts as they address recent social anxieties about increasing levels of antibiotic resistance and concern for developing sustainable agriculture. It repeatedly stresses its role as a training institute and the intimate connection between basic and applied science. The brochure anticipates common criticisms of university-industry relations by stressing that the information produced by researchers involved with the program will be broadly communicated. Interested non-scientists will have access to Institute produced knowledge through the organization's annual meeting and other communication efforts. In addition, "all research generated by the Institute will be published" (College of Agricultural and Life Sciences 1997, 3).

Importantly, the brochure stresses the Institute's commitment to "the protection of intellectual property"(College of Agricultural and Life Sciences 1997, 3). There is a discussion of University policy on intellectual property protection and property rights as well as a discussion of WARF and the organization's role in seeking intellectual property protection. This, as should be clear from the discussion in chapter 5, is crucially important in the Institute's effort to attract industrial support. Indeed, this document is at least in part an effort to generate industry and other support for the Institute. The entire final page of the brochure describes fundraising and the distribution of funds, and specifies the types of contributions the Institute seeks.

This effort is of a very different variety than the bioregional biocontrol collaboration. That effort was itself a challenge to a wide range of existing social values and social structures, and consequently, its likelihood of success on a number of levels was low from the outset. By contrast, this project remains within the boundaries of dominant social values and institutional forms. It comes at a time when federal support for university science has stagnated. It is an era in which "academic capitalism" is rewarded. High-profile academic biologists increasingly have close connections with industry. It is an historical moment, furthermore, in which industry is aware of the thin line dividing basic and applied research in biology and the potential profit to be made by joining with university scientists in search of new products.

IPPM literature stresses the Institute's educational mission, but clearly aims to attract corporate interest by offering firms the opportunity to get early access to on cutting-edge research and the chance to obtain profitable and patented biotechnologies. Indeed, if economic support is a measure of corporate interest in the program, data current through January of 1999 suggest industry enthusiasm. Three Institute projects were being supported by agrichemical companies, and funding was pending from major agrichemical players—Dow, Novartis, and Cargill Seeds—for another four projects. Perhaps not surprisingly, one of the Institute-supported projects that researchers have had difficulty funding is "a science shop model for innovative graduate training" that aims to "serve public needs and under-served constituents" (Buttel et al. to Foster, September 9, 1997). Ironically, one of the two nonprofit foundations that declined the opportunity to support this initiative viewed the proposal as "very sound" but concluded that funding the effort was the responsibility of public institutions (Hesterman to Buttel, September 30, 1997) at a time when universities are increasingly market-driven.

While the social environment limited the likelihood of success of the bioregional biocontrol project—social structure had constraining effects—most efforts promoted by the IPPM will find a supportive social environment. The mission of the Institute is generally consistent with dominant social values and structures. But as Handelsman is all too aware, involvement in the IPPM can itself shape scientists' practices. This is her assessment of the situation: "I think it [the Institute] is a very worthy cause . . . , but the concern I have . . . is the issue of looking like—whether we are or not—we are doing industry's bidding. . . . [I]t's who you talk to, spend time with. The scientists here are going to be hearing and talking to all those industry people who talk a certain language and care about a particular set of issues and will be asking a certain kind of question" (interview, February 16, 1996). Handelsman worries about "everyone else" and wonders where they will "get their voice." But she sees clearly the constraining effects of the existing social organization and can find no way around it:

> The bottom line for me is that my job is to educate . . . , and the funding situation is to the point where I am basically saying whatever it takes to fund my graduate students, I will do, and that's my top priority. And I guess for a long time it just seemed like there were other priorities, and my whole involvement with the alfalfa project at the Center for Integrated Ag Systems [the bioregional biocontrol project], and you know all these little interdisciplinary social science-agriculture grants and everything that we tried to get or in some cases did get . . . , just looking back over the last five years or eight years, it's just a distraction from what I am here to do. . . . Part of me really is concerned about this attitude because I think that . . . looking at the social issues around science is so important, and on the other hand, I say it's not fundable. Society . . . apparently doesn't value it. The Wisconsin public says very much they want people to be looking at . . . the public good of the science we do, but nobody's willing to pay for it, and all it does is take my time away from what my real mission is, which is doing the science. (interview, February 16, 1996)[11]

This episode provides an interesting complement to and contrast with the bioregional biocontrol case. It is similar to the earlier instance in that this case too points to the ways in which the larger social environment can shape the practice of science. These instances both highlight the importance of dominant modes of social organization and established social values. In the bioregional biocontrol case, the work of Handelsman and her colleagues challenged a wide array of existing norms; that confrontation lowered the probability that the project would succeed. By contrast, the Institute is fully consistent with existing social trends, and consequently, it has a higher probability of success.[12]

Conclusion

As many readers will be aware, the title of this chapter is a play on the title of a 1983 article written by Bruno Latour, "Give me a Laboratory and I will Raise the World." Although it is undoubtedly true that laboratories and scientists play a role in shaping the world, my aim has been to show what I see as a more important and significant trend: how the work done by scientists and the world they inhabit is shaped in significant ways by the larger (social) environment in which the practice of science is undertaken.

Specifically, my aim in this chapter has been to two-fold. First, I have tried to illustrate how two efforts that involved the Handelsman lab were shaped in various ways by the character of the social environment in which the efforts were undertaken. Second, in providing these portrayals, I have tried to show that contrary to the claims of Michel Callon and others in the science studies field, it is possible and appropriate to analytically separate the social from the technical. Indeed, the cases I have discussed suggest that there are times when what can broadly be referred to as social factors have temporal and explanatory priority over technical factors. Finally, I have provided a different concept of the inventor-sociologist than that advanced by Callon. The actors whom I have profiled are aware that their work is undertaken in a social environment with particular characteristics and that that environment makes certain projects more difficult to undertake and others easier. In other words, being an inventor-sociologist may involve recognition of the social barriers to technological success more than a naive belief that particular social structures can and must be altered.

The arguments I have presented in this chapter rely on a number of assumptions and counterfactual claims. Let me point to a few. First, I suggest that the difficulty that Handelsman and her collaborators had in getting support for the bioregional biocontrol project cannot be explained exclusively in terms of the intrinsic merits of the project. I supported this claim by pointing to Handelsman's record at receiving funding support, as well as plausible alternative explanations for the funding difficulties. Second, in showing that even if sufficient funds were available, and if these funds would have allowed the gathering of reasonably conclusive evidence for the bioregional biocontrol hypothesis, the project might very well have fallen flat. I noted the economic difficulties small firms face in an economy dominated by large firms, and provided evidence to suggest that any effort on part of the Handelsman group to challenge existing intellectual property norms would probably have been met with resistance.

One final observation: The argument advanced by science studies researchers that it is impossible to separate into hierarchical form the social and the technical is typically presented as a critique of the work of social scientists (see Callon 1987, 83). But scientists too often adhere to the position that it is possible to separate the social from the technical. The technical is understood to exist outside the reach of social forces. Science is the priority.

As a good scientist-sociologist, Handelsman herself is quite aware that spending a disproportionate amount of time "talking to industry" can shape her thinking in a way she might wish to avoid. At the same time, in meeting after meeting, interview after interview, she consistently distinguishes discussions about issues like intellectual property from what she refers to as "the science." When a discussion about ordering research materials or securing funding has run its course, she will direct students to "the science." Listening to Handelsman talk, it becomes clear that it is "the science" that she loves. What brightens her eyes and sparks enthusiasm in her voice is understanding a "biological puzzle" or listening to a student outline an "elegant experiment."

But the reality of running a university biology lab does not allow her the luxury of separating "the science" from matters of patenting, funding, and administration that play important parts in her professional life. This is the case to such an extent that when I arranged a meeting with Handelsman to learn about recent developments in the lab, she told me that there were two significant developments to tell me about. The first was the Institute for Pest and Pathogen Management. The second was a hypothesis that emerged out of the microbial ecology project. One of the things that excited Handelsman about the hypothesis is that it is farfetched, but plausible. Handelsman's discussion of these two developments did not allow for any analytical difference between the two. Handelsman's failure to make this separation would suit Callon. "Ah," he might say, "you see we have evidence that the social and the technical cannot be separated." In fact, however, I suspect Handelsman would like nothing better than to be able to wall off "the science" and not concern herself with the kind of social phenomena (funding pressures, intellectual property worries, disciplinary boundary disputes) that distract her. But the social environment in which she works will not allow that. To do "the science" requires funding. To have companies take inventions seriously requires patenting. Not any patent agent will do, since the lab's resource constraints virtually require that it work with WARF.

Thus, the social and the scientific or technical are intimately connected in the Handelsman lab. But they are not inseparable. If the Handelsman lab is representative, understanding the dynamics of university science today demands consideration of the nature of the separation, the character of the relationship between the social and technical: how the larger social world shapes the practice of academic science.

Afterword

Some years have has passed since I completed my fieldwork in the Handelsman lab. Some things have changed. The composition of the Handelsman group is different. Most of the students who were in the lab when I was there have moved on to teaching positions or professional jobs. To the biocontrol and microbial ecology projects that were underway when I was in the lab, Handelsman scientists have added another initiative. In treating the "midgut" of the gypsy moth as an ecosystem, they are trying to understand the microorganisms within it and are asking questions about the relationship between these organisms and their host, the gypsy moth.

But the overall context in which the Handelsman lab and others like it undertake research has changed very little since I started my investigation. I believe my analysis captures some basic features of university biology today—a biology now inextricably intermeshed in the "new knowledge economy." At the most basic level, I made two arguments in this book. First, I contended that understanding university biology today requires attention not only to the direct effects of the world of commerce on academic scientific practice, but also to the influences of indirect factors. Second, I engaged some of the most prominent theoretical strains in science studies and argued that to understand technoscience demands study not only of processes of construction, but also—and perhaps more importantly—of the ways in which the already constructed features of the social world shape practices in the scientific field and beyond. In this context, we should pay attention to structures, organizations, institutions, and power.[1]

Although I cannot provide more than an outline here, let me sketch two sets of suggestions for future research and action that follow from my investigation. On the research front, I believe the science studies field would be invigorated if more work at the center of debate and discussion took the role of technoscience in the new economy as a pivotal object of study. Of course, there is work that does engage this subject matter, but in my view too much science studies scholarship still treats contemporary technoscience as if it took place autonomously.

Second, I believe that science studies could learn a great deal from recent work in organizational analysis, political sociology, and political economy. There is research that treats these literatures, of course. But typically, it seems that scholars who study technoscience from an organizational or political perspective tend not to talk to science studies scholars.[2]

Beyond future scholarly research, my study has a host of implications for policies and practice that should guide the university as the new millennium proceeds. Let me outline a few. First, university administrators and policymakers should work to create research portfolios that are more balanced than those of many American research universities. Decision makers will need to make an effort to correct current patterns of imbalance in funding and to compensate for years of research support that has created some vibrant areas of investigation while stunting the development of other fields with real potential. This will require a longer view than policymakers often take, as the funding of areas like biocontrol research are long term investments.

Second, policymakers and academic administrators should consider establishing what might be termed institutionalized reflexivity—a set of mechanisms that will allow universities to monitor (and therefore protect from) the clear and egregious influences of university-industry collaborations, *and* also to systematically explore, for example, whether the time spent by academic scientists seeking intellectual property protection is compensated for by the return from licensing of inventions. The question of whether patenting truly enhances university scientists' control over their research should be addressed. In this context, my study suggests that for many scholars the return on the time invested seeking intellectual property protection is low, and that patenting by academic scientists does not always enhance their control over their inventions.

Finally, I believe that the failure of intellectual property protection to pay off for a great many universities and academic scientists should lead state and federal governments to support research on the economic costs and benefits of intellectual property protection and to provide economic incentives for companies and individual researchers who keep their inventions in the public domain.

Overall, I worry that universities in the United States are increasingly seen as engines of economic growth and not institutions of higher *education*. Mitigating the impact of the culture of commerce on universities will not be easy. The indirect effects of the commercial world on the practice of academic science are difficult to see and easier to ignore than the direct factors that have been the focus of controversy. But policymakers and administrators must attempt to confront these indirect effects if there is to be any possibility of balancing equity and efficiency while preserving democratic values in the future university.

Notes

1. Impure Cultures

1. The distinction between basic and applied science is used frequently and unreflectively in the science policy literature and by many scientists whom I have met over the years. The terms are problematic, however. As a wide range of literature in science and technology studies shows, the distinction does not inhere in the objects, but in the representation of the objects (see Latour 1987; Grint and Woolgar 1987). The nature of these representations varies; the terms are often used strategically (Kleinman and Solovey 1995; Webster 1989). Although it is likely that more traditional analysts would conclude that essential attributes are what distinguish basic and applied science, such luminaries as Harvey Brooks recognize that "the identification of basic research with lack of applicability or usefulness or with motivation by pure curiosity is a gross oversimplification" (1993, 222; see also Stokes 1997). Social scientists in a more critical tradition, who might point to inherent characteristics as bases for making these distinctions, suggest that in this particular historical juncture, "technoscience makes impossible the separation of science and technology, basic and applied research" (Slaughter and Leslie 1997, 38).

Despite the problematic nature of these terms, I use them without qualification throughout this book. In general, I use the words as they are used by my informants or in documentary sources. This tack is consistent with the approach I take throughout the text. As I suggest later in this chapter, I believe social phenomena are constructed, but if we focus exclusively on construction processes, we will fail to notice the influence exerted by existing entities on social—including scientific—practices. I am interested in these effects. I leave it to others to consider how and why the concepts of basic and applied science are constructed in particular ways at specific times.

2. When Handelsman relates the story of the "discovery" of UW85, she reveals her generosity as well as something about the nature of university biology today. When Handelsman recounted the history of her lab's precious organism to me, I listened for nearly an hour, and she never spoke about her role in the discovery and understanding of the lab's *Bacillus cereus* strain. Handelsman, like many directors of biology laboratories at research universities, spends little time at the laboratory bench. Instead, she deals with e-mail or is on the telephone talking to colleagues; she is meeting with people—students, collaborators, patrons, or patent agents; she is reading or writing (See Knorr Cetina 1999, 223). This does not diminish Handelsman's intellectual influence in the lab; such a division of labor is precisely what makes the perpetuation of labs like hers possible. With little or no time to spend at the lab bench, I expect Handelsman would happily echo cancer researcher Robert Weinberg's assessment of his professional status: "To the extent I've had success in science, it stems in small part from my own scientific intuitions and in larger part from my success in having good people work for me" (quoted in Angier 1988, 24, 25; see also the similar comment by Paul Berg on receipt of his Nobel Prize for chemistry [Traweek 1988, 88]). As early as the late nineteenth century, scientists who ran large laboratories recognized the collective nature of their endeavors. See, for example, Todes (1997).

3. Press coverage, as important as it is for maintaining public support of university science and perhaps attracting corporate interest, adds little to discussions of UW85 as an object of scientific or commercial interest. It does more to create the myth of UW85—part of only background noise in the lab, but probably a factor in Handelsman's high visibility on campus.

When the *Milwaukee Journal* first covered the work of the Handelsman laboratory in 1989, the paper told the story of a "wonder bug": "In the dirty nasty underworld of bacteria lives a super bug that may rid crops of disease, spur them to new heights and free farmers from chemicals that contaminate wells and may cause cancer." As reported in the *Journal*, the tale of UW85 is the story of a researcher and a student. The researcher was raised "not among the silos of the rural Midwest, but amid the skyscrapers of Manhattan." The discovery was reported as the result of "serendipity and a very inquisitive student" (Jones 1989, A23).

Five years later a University of Wisconsin publication narrated this same account under the title "The Hunt for the Wunderbug" (Gallepp 1994). In that same year, the *Christian Science Monitor* began its chronicle this way: "It's the Timex watch of microbes: Boil it, dry it, deprive it of oxygen, and it keeps ticking—fighting plant disease. UW85, recently discovered by a team from the University of Wisconsin at Madison, combats a strain of algae that specialists say is responsible for causing millions of dollars of damage to crops and ornamental plants" (Spotts 1994).

4. It would be unfair to imply that there was absolute consensus on this point. One of the graduate students who addressed the basic/applied distinction in her discussion of why she liked working in the Handelsman lab indicated her unabashed preference for applied research.

5. Although there is now a trend toward the use of the first person in science writing, historically, omission of the first person conveys an aspect of the dominant view in science that an objective method and habit of mind makes it possible to rigidly separate the subject and the object, and that good science demands it (see Ding 1998 and Ward 1996, 1–16). In keeping with this tradition, according to Charles Bazerman (1988), part of the rhetorical force of scientific articles arises from the fact that the actual human construction of these documents is concealed by their grammatical form.

6. Photographs of field research raise interesting questions, since such representations may not reflect the data. Instead, they serve a rhetorical purpose. They give a visual impression that might very well shape the way viewers think about the data that is presented to them.

7. As with everything in the lab, perceptions are not monolithic. Upon reading this paragraph, one former student commented that she "never felt constrained by money."

8. Knorr Cetina suggests the assignment of service responsibilities in academic biology labs is common (1999, 226).

9. PCR is a central tool in contemporary molecular biology. I discuss it at length in chapter 4 and also in chapter 5. For an interesting sociohistory of the development of PCR see Rabinow (1996).

10. Relatively recent work in science studies stresses the craft character of science and the importance of nonformal or tacit knowledge for successful experimental practice. See, for example, Knorr Cetina (1981), Lynch (1985), and Clark and Fujimura (1992).

11. Diana Forsythe (2001) provides some interesting discussions concerning the complications that result when ethnographers study people who possess social power equivalent or superior to them.

12. For my assessment of the laboratory ethnography literature, see chapter 2.

13. Whether it makes sense to think of the world in terms of levels of analysis such as micro and macro is a matter of some debate in science

and technology studies. I outline my position on this and related issues in chapter 2.

14. Upon reading this, one former member of the lab cautioned me not to take Handelsman's reaction as universal. This researcher said that a number of people in the lab simply did not follow my presentation, and commented, "At the outset I hoped biologists would get something out of your book, but after your talk I began to think that biologists reading and understanding your book was about as likely as gorillas reading Diane [*sic*] Fossey's research."

15. Latour and Woolgar touch on these issues (1979, 72).

16. Handelsman and I have not talked about these tensions since, but I have sought and received her advice on another research project, and we are currently coediting a book series for the University of Wisconsin Press. I have also not talked to Goodman about the situation, but again, I have sought and received his advice; we are on good terms.

17. Discussing work of this variety in the history of technology, MacKenzie and Wajcman suggests the work emphasizes the "heroic inventor." According to proponents of this approach, "great inventions occur when in a flash of genius, a radically new idea presents itself almost ready-formed in the inventor's mind. This way of thinking is reinforced by popular histories of technology, in which to each device is attached a precise date and a particular man (few indeed are the women in such lists) to whom the inspired invention belongs" (MacKenzie and Wajcman 1985, 9).

18. Whether my analysis of a similar lab outside the United States would have produced the same reaction by the lab leader, I cannot say. Insofar as characteristics of American culture shaped Handelsman's response, I can imagine the leader of a lab outside of the United States might very well have reacted differently.

Of course, it is possible that Handelsman's reaction was not shaped by the culture of individualism found in the United States. It may simply be, as Brian Martin suggested to me, that scientists are used to being the scrutinizers, not the scrutinized, and so are uncomfortable being studied. A still different explanation, as Martin also suggested to me, comes from social psychology. This is the notion of a fundamental error of attribution. This error occurs when one explains the behavior of others in terms of larger social or psychological variables, but attributes her or his own reactions to highly contingent or context specific factors.

19. For an interesting discussion of people's failure to grasp the larger structures affecting their lives, see Eliasoph (1998).

20. I develop my understanding of structure in chapter 2.

2. Traversing the Conceptual Terrain

1. Portions of this section of chapter 2 appeared originally in Kleinman and Vallas (2001). I am grateful to the publisher (Kluwer Academic Publishers) of *Theory and Society* for permission to reprint this material.

2. Although there is a global trend toward university involvement in the world of commerce, my discussion here concerns only the United States. For work on university-industry relations outside the United States see, among other studies, Katherine Balazs and Guilhelrme Ary Plonski, Academy-Industry Relations in Middle-Income Countries: Eastern Europe and Ibero-America, in *Capitalizing Knowledge: New Intersections of Industry and Academia,* edited by Henry Etzkowitz, Andrew Webster, and Peter Healey (Albany, N.Y.: SUNY Press, 1998), 151–168; Mike Berry and Lioudmila Pipiia, Academic-Industry Relations in Russia: The Road to the Market, in Etzkowitz, Webster, and Healy (1998), 169–186: Margaret Sheen, Universities in Scotland and Organizational Innovation in the Commercialization of Knowledge, in Etzkowitz,Webster, and Healey (1998), 187–214; Magnus Gulbrandsen, Universities and Industrial Competitive Advantage, in *Universities and the Global Knowledge Economy: A Triple Helix of University-Industry-Government Relations,* edited by in Henry Etzkowitz and Loet Leydesdorff (London: Pintor, 1997), 121–131; Morris Low, Japan: From Technology to Science Policy, in Etzkowitz and Leydesdorff (1997), 132–140; Kathryn Packer. 1995. The Role of Patenting in Academic-Industry Links in the UK: Fool's Gold? *Industry and Higher Education* 9 (5): 293–302; Andrew Webster and Julian Constable. 1990. Strategic Research Alliances and Hybrid Coalitions. *Industry and Higher Education* 4: 225–230.

3. It is worth noting that establishment of such programs was not always smooth or completely successful from the perspectives of participants, and many of the issues raised in recent years about the compromise of the university's mission and the need for administrative oversight of university-industry relations were taken up by parties involved in discussions about establishing early twentieth century programs to permit business-academic collaboration. On this topic, see Servos (1996).

4. Geiger notes: "DoD domination of university research, insofar as it existed, was always concentrated in specific departments and specialties. In aeronautical engineering, parts of electrical engineering, optics, underwater acoustics, and numerous other specialties the DoD consistently supported research that would not otherwise have been performed. This kind of support cemented long-term relationships and a

degree of influence that substantially affected units concerned" (1993, 194, 195). DoD support was also focused on particular universities, and Stanford and MIT were two of the three top recipients of DoD funds during the early Cold War years. (In addition to the work drawn on for my discussion of military research at Stanford and MIT, see Leslie [1993] and Noble [1977].)

5. It is worth mentioning again that the distinction between basic and applied research is unclear and generally problematic. The terms do not reflect something intrinsic to the research, and we are justified in being skeptical of these departments' "admissions" that their research shifted to applied work. It is quite possible that the substantive character of the work changed very little under these circumstances, but that new patrons required that scientists represent their work differently. During its first decade, the National Science Foundation used the terms *basic* and *applied research* as a means to protect what administrators saw as the agency's objectives (Kleinman and Solovey 1995).

6. Well over 20 percent of these centers were founded between 1960 and 1980. This is important to point out lest we fall into the habit of mythologizing the period before university-industry relations in the biological sciences.

7. David Hess urges caution in calling this work ethnographic, since "there is little if any thick description or semiotic analysis of local categories, contradictions, and complexities; there is little sense of cultivation of informants, talking to people, finding out what they think, understanding their social relations, and analyzing the play of similarity and difference across domains of discourse and practice. In short, there is little if any culture. What tends to happen instead is that the sociological theories and (anti)philosophical arguments upstage the stories and worlds of the informants" (1997b, 156).

8. Michael Lynch makes a clever observation in this context: if a fact is simply a statement with no modality or trace of authorship, since this statement includes no modalities, he concludes that "it is as good an example as any of a fact" (Lynch 1993, 93).

9. Latour and Woolgar suggest that to make this claim is not a denial of the solidity of facts, but only an emphasis on the context in which facts are created. At the same time, Latour and Woolgar's position is not realist in the sense that they argue that "'reality' cannot be used to explain why a statement becomes a fact, since it is only after it has become a fact that the effect of reality is obtained" (1986 [1979], 180).

10. As provocative as Latour and Woolgar's study is, Lynch argues that the work does not demonstrate that facts are constructed, since the authors begin by assuming this. Instead, says Lynch, "It would be more accurate to say that they demonstrate that a constructivist vocabulary

can be used for writing detailed descriptions of scientific activities" (1993, 102).

11. Despite her more macro-cultural orientation, I have grouped Traweek with the classic laboratory studies because this is the way others commonly classify her. David Hess suggested to me that in many ways Traweek's work is better placed among other ethnographies of science done by anthropologists. Work by analysts like Emily Martin (1990) links the laboratory with the world beyond it. Other work by anthropologists of science and technology takes up macro-structural concerns (see Downey 1998) and theories of power (see Hess 1995b). These are concerns of mine as well. Other interesting scholarship by anthropologists working in science and technology studies can be found in Downey and Dumit (1997). See also Forsythe (2001). Finally, an often overlooked study that places the laboratory investigated in a larger context is Charlesworth et al. (1989).

12. In his analytical review of ethnographies in science and technology studies, Hess (2001) suggests that, unlike many first generation ethnographies in STS, work in the second generation tends to focus more on social problems, is likely to be "multi-sited" (in my case looking at the relationship between the lab and other social spaces and social structures), and tends to make concepts like power and culture more central to the analysis.

13. I accept Latour and Woolgar's assessment in their 1986 postscript to *Laboratory Life* that although in the end "the laboratory should not be studied as an isolated unit" (1986, 280), such isolation was probably necessary in the early laboratory ethnographies.

14. To be fair, Latour and Woolgar are sometimes attentive to the impact of such factors as the political situation at the level of the state (1979, 123) and disciplinary status (1979, 146). But concern with such features is by no means consistent.

15. On this and related matters, see Jason Owen-Smith (2001).

16. For related approaches to power in science and technology studies, see Downey (1998) and Hess (1995b).

17. For some recent actor network–related work, see Law and Hassard (1999).

18. These two orientations in no way exhaust the range of perspectives that constitute the field of STS. Among other recent and interesting perspectives, which I do not explicitly discuss, are the mangle approach championed by Andrew Pickering (1993; 1995) and post-essentialist research promoted by Steve Woolgar and his colleagues (see Grint and Woolgar 1997)

19. David Hess (1997, 103) points out that although laboratory studies had largely disappeared from the scene by the 1990s, "their legacy of

emphasizing what scientists do was continued in the 1990s in research" under the science-as-practice rubric, including the social worlds approach and the mangle framework advocated by Pickering (1993; 1995).

20. I am generalizing here, and not every aspect of my critique applies to everything written in these two traditions. On the matter of constraint, for example, see Callon (1991).

21. For a related criticism, see Hess (1995b).

22. For a related critique of what she refers to as the action framework, see Knorr Cetina (1999, 9).

23. The range of work in science and technology studies that might broadly be grouped under this rubric is extensive. See, among other scholarship, work by Epstein (1996), Hess (1995b), Klein and Kleinman (2002), Kleinman (1995), B. Martin (1988; 1999), Moore (1996), Rappert and Webster (1997), Restivo (1995), Webster (1994b,) and Wright (1994). Although I place this work in a single category, there is significant variation within this body of scholarship; it is quite possible some of the authors I cite might reject my characterization.

24. My work is not the first attempt to conceptually balance structure and agency. Among the most prominent theorists who seek to circumvent the structure-agency dichotomy is Anthony Giddens (see 1984). Giddens provides an analysis of the ways in which structures are carried by knowledgeable actors and the ways in which structures are reproduced in and through everyday practices. Within the context of a general theory of the relationship between structure and agency, Giddens at once takes seriously the idea that structures are constructed and that at any given moment they can be stable and constraining.

My conceptual objective in this study is far more modest than Giddens's project. How the balance of structure and agency works and how social order is reproduced is *not* my concern here. Instead, I aim to take a synchronic look at several distinctive structures and to illustrate the ways in which these structures can constrain or shape actors' practices.

Two interesting efforts to use Giddens's theory of structuration in science studies are Abir-Am (1987) and Hagendijk (1990). For a thought-provoking effort to grapple with the structure-agency problem in different empirical terrain see Fine (1996).

25. In a comment related to my critique of agency-centered approaches and many of the laboratory studies, David Hess contends that "the great macro-sociological issues opened up by the Edinburgh school—class, the state, race, gender, colonialism, historical transformations—were largely irrelevant for a kind of analysis that focused on the making of science in small groups, networks, and shifting institutional fields. As a result, the critical potential of SSK was limited" (Hess 1997, 105).

3. Braided Paths

1. The synthetic organic herbicide 2,4-D was introduced in 1944, the year before DDT. It is typically highlighted in histories of the war and the development of agricultural chemicals (Harrington 1996, 415).

2. A wide range of institutional analysts argue that large scale crises, such as war or economic depression, provide an environment in which major institutional change is likely. See, among other studies, Hall (1986), Kleinman (1995), Krasner (1984), and Skowronek (1982).

3. Measured by active ingredient, Osteen and Szmedra suggest that this figure was much less, nearly 900 million pounds in 1982 (1989, 10).

4. Commentators who make similar points include Odrish (1976, 179) and Perkins (1982, 11).

5. It is worth noting that the absolute number of biological control papers cited in the *Review* did not decline after the War.

6. Exploration for and acquisition of this biological material raises important questions about rights of access and ownership of such substances. Controversy around these issues increased after recent developments in recombinant DNA technology. See Kloppenburg (2000; 1990), Kloppenburg and Kleinman (1985), and Shiva (1997).

7. In her first laboratory ethnography, Knorr Cetina makes a related point that was apparently lost on most practitioners in later science studies: "The scientists' laboratory selections constantly refer us to a contextuality beyond the immediate site of action" (Knorr Cetina 1981, 81).

4. (Un)Intended Consequences

1. The contributions to this volume represent a diversity of perspectives.

2. Gaudilliere and Lowy (1998a) make a related point suggesting that the case studies in their edited volume show how standardized research instruments and homogeneous research materials permit the stabilization and diffusion of locally produced knowledge.

3. In more recent work, analysts take similar positions, focusing on the advantages of standardizing and black boxes (see Knorr Cetina 1999, 84; Shinn 1997, 86).

4. For interesting exceptions, see, for example, Webster and Packer (1997) and work in Gaudilliere and Lowy's edited collection (1998c).

5. Of course, Latour and Woolgar treat black boxes as bases of power asymmetries in the process of fact construction where the creators of black boxes have power over prospective challengers who face substantial costs in any effort to open black boxes.

Among the early laboratory ethnographies, Sharon Traweek's *Beamtimes and Life Times* is exceptional in its attention to the power embedded in relationships around research tools. Traweek considers, for example, the powerful position of resident experimentalists vis-à-vis visitors at high energy physics facilities and the powerful position of resident tool building engineers vis-à-vis experimentalists (1988, 35, 36). Traweek is attentive, furthermore, to that most important resource—money—in shaping research practice (1988, 50).

6. This orientation sets my approach apart from recent work like Fujimura's (1996). While Fujimura is interested in "how and why these [molecular biological] techniques *came to be* the keys to research on human cancer" (1996, 70; emphasis added), I am interested in the impact of particular technologies after *they have already become* routine parts of laboratory work.

7. Although I know of no work that takes precisely this perspective, there is certainly work that explores different ways in which standardized research tools constrain research practice. Thus, for example, work in Gaudilliere and Lowy's collection looks at the ways in which standardized tools shape theoretical choices (1998c) and work by Clarke (1987), Oudshoorn (1990), and others looks at how research materials shape social and cognitive developments in science. Gaudilliere and Lowy (1998b) also consider how uniform research mice lead to uniformity in research practices.

8. For many years, research tools and related infrastructure were largely ignored by social analysts of science (Clarke 1987, 323). There is now, however, a growing literature on the socio-history of these materials. See, for example, the collection edited by Clarke and Fujimura (1992) and Bud and Cozzen's edited volume (1992). In addition, see Robert Kohler's work on drosophila (1994), Clarke's essay on physiology (1987), Bud and Warner's encyclopedia of scientific instruments (1998), and the Gaudilliere and Lowy collection (1998c).

9. On the importance of customized tools in high energy physics, see Traweek (1988) and Knorr Cetina (1999).

10. In this connection, Clarke (1987, 338) notes: "The high costs of live materials drew biomedical scientists into new relations with external funding sources and into the arena we now call 'big science.'"

11. Some academic labs do have the capacity to undertake fatty acid analysis, and they might do tests for other academic labs.

12. Similar problems are not unheard of in the lab. One Ph.D. student told me of how she ran into a such a problem once when she purchased a restriction enzyme. The enzyme (and the company-supplied control) did not work as it was supposed to. When the student spoke to a company representative, the official was unwilling to acknowledge

that the source of the problem could have been with the company-supplied enzyme rather than with the scientists' use of it.

13. Or they must find a university lab that has the equipment and is willing to undertake the test, as Handelsman scientists were ultimately able to do.

14. Of course, the company believed the test is reliable.

15. It is important to note that in this instance members of the Handelsman lab and the company staff did not disagree on what constituted accuracy. Instead, they disagreed about the likely source of inaccuracy.

16. As I discuss in the next chapter, intellectual property law can impede efforts at such customization.

17. A former student in Handelsman's lab took issue with Handelsman's assessment of restriction enzymes, and of the Handelsman staff scientist's analysis of the Southern blot episode. He said, "I think to an extent every generation looks at the new generation and is aghast at the things they take for granted and don't know in depth. I also think that some students don't understand the underlying principle of what they are doing regardless of whether it is a kit, something from another lab, or just an in-house protocol they are following" (correspondence May 19, 2001).

18. The National Institutes of Health is a government agency, not a commercial concern. Consequently, as I note below, this case points more to the ignorance that can result from standardization and black boxes than to a problem produced by commercialization. And, as I suggested at the outset of this chapter, standardization is propelled not just by commercial factors, but by the drive for increased efficiency among scientists. At the same time, many databases available to molecular biologists are commercially produced and maintained. These data sources themselves are a source of different concerns.

19. Although Handelsman spoke of these problems on several occasions during my six months in her laboratory, three years later when I asked her to discuss these in detail, she had difficulty providing concrete examples. After talking with her graduate students, she concluded that perhaps the problem was not as serious as she initially thought. Indeed, she was led to conclude that in some instances not only did kits *not* hinder advanced education, but for the self-directed student, they could actually facilitate training. The "good kits," Handelsman told me, provide detailed instructions, and these may actually teach careful readers about biological and chemical processes.

Beyond the substance of Handelsman's change of perspective, it is interesting to consider how my presence and probing may have led some lab workers to examine their perspectives and practices. This is one of the complications and possible benefits of ethnographic research. Also, it is not only lab workers' orientations that may be altered. As I note in the prologue, my opinion of laboratory science and scientists was transformed by my time in the lab.

5. Owning Science

1. On the political economy of bioprospecting, see Kloppenburg (2000).

2. For more recent information on this controversy, see Martin Enserink. 2000. Start-Up Claims Piece of Iceland's Gene Pie. *Science* 287 (February 11): 951 and Martin Enserink. 1999. Iceland Oks Private Health Databank. *Science* 283 (January 1): 13.

3. For a slightly different approach to the increasing restrictions on the flow of information in the biological sciences, see MacKenzie, Keating, and Cambrosio (1990).

4. This brief literature review barely scratches the surface. The work on intellectual property and the biological sciences is extensive. Scholarship by Etzkowitz, Packer, and Webster straddles the policy literature and the theoretical terrain of science studies (see Packer and Webster 1995; Etzkowitz and Webster 1995). On the patenting of genomic information see, for example, Heller and Eisenberg (1998) and Hilgartner (1998). Geof Bowker provides a broadly constructivist account of patenting and patent struggles in the petroleum industry (1995). For an example of work on debates on copyright issues see, for example, Herrington (1998; 1999). For more fundamental critiques of the logic of intellectual property protection, see Boyle (1996), Martin (1998), Shulman (1999), and Vaidhyanthan (2001). For a different take on the public and private in biotechnology, see Gieryn (1998).

5. For a compelling and collegial critique of Myers essay, see Packer and Webster (1995).

6. This literature is discussed in the introduction to this chapter as well as in the sections on the norms of science and in the chapter's final section about WARF.

7. I do not want to exaggerate the breadth of the "experimental use exemption." First, this exemption to patent infringement is not sanctified in statute. It is recognized in common law—that is, in the history of court decisions—but only in a narrow sense. Traditionally, the exemption holds only for research of the most basic sort and does not apply to scientific work where there is any commercial aim. For further discussion of the experimental use exemption, see Eisenberg (1989), Mueller (2001), and Parker (1994).

8. In recent years, the National Park Service has sought to make a return on the work of bioprospectors in the national parks. After the NPS signed a four year deal with a San Diego biotech company worth $175,000 plus up to 10 percent of future sales for bioprospecting in Yellowstone, the Park Service was sued by three nonprofit organizations aiming to halt the agreement. The suit asserted that the bioprospecting agreement violates the law that requires that national parks be preserved in an unspoiled state, and that also requires environmental impact assessments on proposed projects within national parks (Pennisi 1998). In March of 1999, the bioprospecting deal was halted by a federal court judge until possible environmental impacts could be determined (Lambrecht 1999, A1), but the judge ultimately dismissed the suit just over a year later. At the time, the plaintiffs said they would appeal (Pollack 2000, F5), but they have since withdrawn their appeal.

9. In a 1999 decision, a federal judge ruled that Roche's patent on the native *Taq* enzyme is not valid. This decision was appealed. Perhaps more significantly, however, Roche, has a patent on a recombinant version of *Taq,* and it is this version that now dominates the *Taq* market (Service 1999).

10. Roche has announced that it does not intend to pursue researchers doing "pure science," but the company is keeping tabs on these "patent infringers," nonetheless.

11. Such challenges to the experimental use exemption are likely to become more common in the years ahead. Packer and Webster note, "the more that universities take on a commercial function, the more that their research might be regarded as less than purely 'experimental' and the less likely they will be seen as disinterested users. Patent holders—indeed, other universities—may become much more cautious about allowing exemption to research teams in universities whom they know are likely to be encouraged to commercialize any new findings that emerge from their work" (1997, 54).

12. A former lab member commented to me that Handelsman and Goodman's discussion of manufacturing *Taq* in-house is only "slightly more realistic than the students talking about putting in a soft-serve ice cream machine" (correspondence, May 19, 2001).

13. After reading this discussion, a former student in the lab commented: "I agree that LaRoche has lots of control . . . to change the rules" (correspondence, August 27, 2001).

14. A similar situation arose a few years back with the so-called cystic fibrosis gene. This gene and the principle mutation associated with cystic fibrosis were patented by the universities of Toronto and Michigan, and these institutions expressed an inclination to require researchers to

pay royalties for using this genetic material (*Biotechnology Business News* 1993).

15. Gardner and Rosenbaum (1998) describe a hypothetical case of database construction that also raises issues similar to those brought into view by the *Taq* case.

16. Disturbed by the restrictions imposed by some MTAs, the National Institutes of Health established guidelines that prohibit scientists who receive federal support from signing MTAs that require withholding data or give providers of research materials a property claim on discoveries from use of those materials. Government funded scientists may similarly not seek or agree to exclusive licenses on research tools (Enserink 1999; Marshall 1999).

It is also worth pointing out that the trend toward what might broadly be termed the commodification of 21st century biology is not linear. Beyond this protective action by the NIH, in this chapter I describe the Patent and Trademark Office's raising of the threshold that must be met for would be inventors of biologically based technologies to secure patent protection. Also, recently the Wisconsin Alumni Research Foundation set up a nonprofit organization for the distribution for research purposes of embryonic stem cells (Vogel 2000).

17. A recent case involving pregnancy screening does not fit with those I have discussed in that it does not involve academic sciences, but it bears enough similarity to these cases to be worthy of mention. In 1986, Mark Bogart sought a patent for a test based on his discovery that levels of frequently checked hormones in pregnant women can provide an indication of likely congenital birth defects—Downs Syndrome, in particular. He was awarded the patent three years later. On the basis of this patent, Bogart's company requires diagnostic laboratories to pay large royalties when they conduct tests like Bogart's. The company has threatened legal action against labs that use the test without payment. Workers in the area of prenatal testing are concerned that the increased cost for use of the test will lead them to discontinue use of it (Eichenwald 1997, A1, C3).

A recent effort by Human Genome Sciences to patent the meningitis bacteria raises similar issues (King and Brown 1998, 1, 3). Clinicians and researchers fear that such patenting could make any immunization developed prohibitively expensive.

At some basic level these cases differ from the others I have considered because they do not involve research as such, and the tool at issue is not used by academic scientists. What they share, however, is the presence of a resource dependence relationship for use of an important tool, as well as the fact that cost could prevent the tool's use. Furthermore, despite legislation preventing situations like this in the future, clinicians are unlikely to be in a position to flaunt the law, since they likely lack the capacity to pursue a legal challenge.

18. If one views Merton's norms of science as elements of an ideal type (see Weber 1949), then perhaps these criticisms are unfair. In fact, some historical research suggests that there are cases where the "moral economy" underlying the exchange of research materials looks a lot like Merton's representation of communism (Kohler 1994). At the same time, it is unreasonable to assume that scientific communism operated virtually transparently and without variation due to context throughout the history of academic science until the biotechnology revolution.

19. Even in high energy physics where commercial motivations are typically nonexistent, "Groups are very careful to manage the information that leaves their group" (Traweek 1988, 114). And in two recent cases, graduate students in education and mathematics—fields not usually known to produce marketable commodities—sued their professors for misappropriating their research (Marshall 1999).

20. If the situation in the United States bears any similarity to that in the United Kingdom, prospects for huge windfalls from academic patenting seem small indeed. In their study of the U.K. scene, Webster and Packer found only four universities that made over £50,000 annually from their licensing activities (Webster and Packer 1997, 49).

21. Handelsman stresses that although WARF's "success rate" is low, the organization's patenting and related investment efforts provide a substantial sum to the University of Wisconsin: between about $16 and 18 million annually. This record, however, makes the University of Wisconsin unusual. According to Geiger, just four universities—Stanford, MIT, California, and Wisconsin—received over half of all American university income from patents in the recent past (1993, 318). According to Nelson, most university licensing offices "barely break even" (1998, 1460). On the other hand, the profitability of patenting of biological inventions undertaken in the 1980s and 1990s will not be clear for years to come. On this issue, see also Slaughter and Leslie (1997, 202).

22. Webster and Packer point to an important irony in regard to the realization of increased control that academic scientists hope to achieve through patenting. They note that any patent claim requires scientists to engage in "reduction to practice" experimental work. This work may be tedious and take many months. Thus, the decision to proceed with patenting distributes intellectual labor and attention in a way that is different than how scientists might proceed were they to forego patenting (Webster and Packer 1997, 53).

23. In the mid-1990s, WARF spent about $17,000 to file, prosecute, and have issued each U.S. patent.

24. The cost in the United Kingdom is considerably less, according to Webster and Packer (1997, 54). File, search, and examination by the U.K. patent office costs about three hundred pounds. Securing the services of a patent agent adds another thousand pounds. Still, one university department in the Webster and Packer study accumulated a £60,000 debt while engaging in patenting activities. The apparently low cost of filing in the United Kingdom does not include any expenses scientists might incur if they need to litigate to protect their patent.

25. This case is considered in detail in chapter 6.

6. It Takes More than a Laboratory

1. As I discuss at length in chapter 2, this is the position taken by actor-network theorists and social-worlds analysts.

2. ATFFI is now called the Program on Agricultural Technology Studies (PATS).

3. Monsanto was a leading independent agri-chemical company until its merger with Pharmacia Corporation in April 2000. It is a mass production organization generally interested in selling a single product to the largest possible market. Under most circumstances the company would have little interest in customizing pesticides to fit the needs of individual farms or regions.

4. To suggest that "social" factors led Handelsman to abandon the bioregional biocontrol project is not to argue that had researchers undertaken further work on the project they would have avoided "technical" difficulties. Handelsman noted that acquiring statistical understanding about variation in biocontrol agent performance linked to region of discovery/region of use would always be difficult, since disease pressure would be the greatest single variable and would vary significantly with site.

5. There is an extensive literature that challenges traditional views on the objective and meritocractic character of peer review. See, for example, Chubin and Hackett (1990).

6. The widely accepted view today is that the rigidity of large-scale enterprise means its days are numbered, soon to be replaced by smaller, more flexible firms (Piore and Sabel 1984). This perspective has been forcefully challenged by Bennett Harrison (1994).

7. On concentration in the commodity processing sector, see Heffernan (2000).

8. When traditional agri-chemical companies do develop and market biopesticides, the model they utilize for introduction and use is not likely to be sustainable. In synthetic pesticide development, producers have confronted a "pesticide treadmill" where widespread use has led

to the development of pest resistence. Thus, widespread blanket use of biopesticides is likely to lead to the emergence of pest resistence, and the biological effectiveness of these agents is likely to be short-lived. See Robin Jenkins. 1998. BT in the Hot Seat. *Seedling* 15 (3): 13–21.

9. I do not want to suggest that major corporations never produce for niche markets. Of course, it is the case that in addition to the dominant forms of organization and production in a given sector there are nondominant and often competing forms.

10. In keeping with current trends (Harrison 1994), a large firm could create a decentralized product development and distribution system for bioregionally based biocontrol products. It is not entirely clear, however, that such a scenario would meet the rural development objectives of Handelsman and her collaborators. Furthermore, a large agri-chemical concern would need an incentive to move away from the mass production of more traditional pesticides. It is not clear what kind of pressure would prompt such a move.

11. In thinking about her varied efforts, Handelsman develops an almost elite versus mass evaluation, which suggests that dominant actors will get what they want, while the masses unable or unwilling to provide financial support for scientific research will be left unattended to.

12. My predictions notwithstanding, the IPPM did ultimately fold. According to one administrator, however, it failed because nobody on campus was willing to take leadership of the Institute. A different picture emerges from the account of an involved faculty member. That scientist indicated that corporate funding for the Institute was not a problem. Indeed, shortly before the Institute was terminated, Novartis provided a quarter of a million dollars to support student fellowships. However, when a new administration came into the College of Agricultural and Life Sciences, the leadership was not interested in IPPM. They cut university funds for the program and gave it no visibility.

Afterword

1. Culture too requires attention in our efforts to understand science and the new economy. See nn.7, 11, chapter 2. See also Kleinman and Vallas (2001).

2. See, for example, Hage and Hollingsworth (2000), Powell (1998), Saxenian (1994), and Slaughter and Leslie (1997).

Bibliography

Abir-Am, Pnina. 1987. The Assessment of Interdisciplinary Research in the 1930s: The Rockefeller Foundation and Physico-Chemical Morphology. *Minerva* 26(2):153–176.

———. 1987. The Biotheoretical Gathering, Trans-Disciplinary Authority and the Incipient Legitimation of Molecular Biology in the 1930s: New Perspective on the Historical Sociology of Science. *History of Science* 25:1–70.

Alexander, Graham. 1994. A Haystack of Needles: Applying the Polymerase Chain Reaction. *Chemistry and Industry* 18:718.

Amsterdamska, Olga. 1990. Surely You Are Joking, Monsieur Latour! *Science, Technology, and Human Values* 15:495–504.

Angier, Natalie. 1988. *Natural Obsessions: The Search for the Oncogene*. Boston: Houghton Mifflin Company.

Arthur, Charles. 2000. High Stakes Battle Is Joined in Deciding Who Owns Our Genes. *The Independent* (London), March 16, p. 21.

Axt, Richard G. 1952. *The Federal Government and the Financing of Higher Education*. New York: Columbia University Press.

Bacrach, Steven, R., Stephen Berry, Martin Blume, Thomas von Foerster, Alexander Fowler, Paul Ginsparg, Stephen Heller, Neil Kestner, Andrew Odlyzko, Ann Okerson, Ron Wigington, Anne Moffat. 1998. Who Should Own Scientific Papers? *Science* 281 (September 4): 1459–1460.

Bagla, Pallava. 1999. Model Indian Deal Generates Payments. *Science* 283 (March 12): 1614, 1615.

Barrell, Bart. 1991. DNA Sequencing: Present Limitations and Prospects for the Future. *The FASEB Journal* 5(1):40–45.

Baron Consulting Company. www.baronconsulting.com/fsl.htm. Taken from website, October 9, 1997.

BBI Newsletter. 1995. Market and Technology Updates: Patent Battle Casts Wide Net. *BBI Newsletter* (June) 6:18. Taken from Nexus.

Bechtel, William. 1986. Introduction: The Nature of Scientific Integration. Pages 3–52 in *Science and Philosophy: Integrating Scientific Disciplines*, edited by William Bechtel, Dordrecht, Netherlands: Martinus Nijhoff Publishers.

Becker, J. Ole and Franz J. Schwinn. 1993. Control of Soil-borne Pathogens with Living Bacteria and Fungi: Status and Outlook. *Pesticide Science* 37: 355–363.

Behling, Ann. 1993. New Disease Fighter Comes from Alfalfa: UW85 Combats Phytophthora, Pythium in Midwestern Soybeans. *Soybean Digest* (December): 34.

Biotechnology Business News. 1993. US Demands Patent Royalties. *Biotechnology Business News*, January 29. Taken from Nexus.

Blumenstyk, Goldie. 1999. How One University Pursued Profit from Science—and Won. *The Chronicle of Higher Education*, February 12, pp. A39, A40.

Blumenthal, David, Sherrie Epstein, and James Maxwell. 1986. Commercializing University Research: Lessons from the Experience of the Wisconsin Alumni Research Foundation. *New England Journal of Medicine* 314(25):1621–1626.

Blumenthal, D., M. Gluck, K. Louis, A. Stoto, and D. Wise. 1986. University-Industry Research Relations in Biotechnology: Implications for the University. *Science* 232 (June 13):1361–1366.

Bowker, Geof. 1992. What's in a Patent? Pages 53–74 in *Shaping Technology/Building Society: Studies in Sociotechnical Change*, edited by Wiebe E. Bijker and John Law. Cambridge, Mass.: MIT Press.

Boyle, James. 1996. *Shamans, Software, and Spleens: Law and the Construction of the Information Society*. Cambridge, Mass: Harvard University Press.

Brooks, Harvey. 1993. Research Universities and the Social Contract for Science. Pages 202–234 in *Empowering Technology: Implementing a U.S. Strategy*, edited by Lewis M. Branscomb. Cambridge, Mass.: MIT Press.

Brown, Paul, and Sarah Boseley. 1998a. Super Bug Threat to Health. *The Guardian* [London and Manchester]. April 23: 1.

———. 1998b. Medicine's Over-Performed Miracle. *The Guardian* [London and Manchester]. April 23: 5.

Brown, Phil and Edwin J. Mikkelsen. 1990. *No Safe Place: Toxic Waste, Leukemia, and Community Action*. Berkeley: University of California Press.

Bucciarelli, Louis L. 1996. *Designing Engineers*. Cambridge, Mass.: MIT Press.

Bud, Robert, and Susan Cozzens. 1992. *Invisible Connections: Instruments, Institutions, and Science*. Bellingham, Wash.: International Society for Optical Engineering.

Bud, Robert, and Deborah Jean Warner, eds. 1998. *Instruments of Science: An Historical Encyclopedia*. London: Garland Publishing.

Bugos, Glenn E., and Daniel Kevles. 1992. Plants as Intellectual Property: American Practice, Law, and Policy in World Context. *Osiris* 2nd Series (7):75–104.

Burawoy, Michael. 1991a. Introduction. Pages 1–7 in *Ethnography Unbound: Power and Resistance in the Modern Metropolis,* Michael Burawoy et al. Berkeley: University of California Press.

———. 1991b. The Extended Case Method. Pages 271–287 in *Ethnography Unbound: Power and Resistance in the Modern Metropolis,* Michael Burawoy et al. Berkeley: University of California Press.

Busch, Lawrence, William B. Lacy, Jeffrey Burkhardt, and Laura R. Lacy. 1991. *Plants, Power, and Profit: Social, Economic, and Ethical Consequences of the New Biotechnologies*. Cambridge, Mass.: Blackwell Publishers.

Business Wire. 1993. Hoffman-La Roche and Perkin-Elmer Announce License Grant to United States Biochemical. *Business Wire,* December 1. Taken from Nexus.

Buttel, Frederick and Jill Belsky. 1987. Biotechnology, Plant Breeding, and Intellectual Property: Social and Ethical Dimensions. *Science, Technology, and Human Values* 12(1):31–49.

Buttel, Frederick, Martin Kenney, Jack Kloppenburg, Douglas Smith, and J. T. Cowan. 1986. Industry/Land Grant University Relationships in Transition. Pages 296–312 in *The Agricultural Scientific Enterprise: A System in Transition,* edited by Lawrence Busch and William B. Lacy. Boulder, Colo.: Westview Press.

Callan, Nancy W., D. E. Mathre, James Miller. 1990. Bio-priming Seed Treatment for Biological Control of *Pythium ultimum* Preemergence Damping-off in *sh2* Sweet Corn. *Plant Disease* 74(5): 368–372.

Callon, Michel. 1995. Four Models for the Dynamics of Science. Pages 29–63 in *Handbook of Science and Technology Studies,* edited by Sheila Jasanoff, Gerald E. Markle, James C. Petersen, and Trevor Pinch. Thousand Oaks, Calif.: Sage Publications.

———. 1991. Technico-economic networks and irreversibility. Pages 132–161 in *A Sociology of Monsters: Essays on Power, Technology, and Domination,* edited by John Law. London: Routledge.

———. 1986. Some Elements of a Sociology of Translation: Domestication of the Scallops and the Fishermen of St Brieuc Bay. Pages 196–233 in *Power, Action, and Belief: A New Sociology of Knowledge,*

Sociological Review Monograph 32, edited by John Law. London: Routledge and Kegan Paul.

Callon, Michel, and John Law. 1989. On the Construction of Sociotechnical Networks: Content and Context Revisited. *Knowledge and Society* 8:57–83.

Caltagirone, L.E., and R.L. Doutt. 1989. The History of the Vedalia Beetle Importation to California and Its Impact on the Development of Biological Control. *Annual Review of Entomology* 34:1–16.

Cambrosio, Alberto, Camile Limoges, and Denyse Pronovost. 1991. Analyzing Science Policy-Making: Political Ontology or Ethnography. *Social Studies of Science* 4:775–781.

Cambrosio, Alberto, Peter Keating, and Michael Mackenzie. 1990. Scientific Practice in the Courtroom: The Construction of Sociotechnical Identities in a Biotechnology Patent Dispute. *Social Problems* 37: 301–319.

Campbell, Donald T. 1969. Ethnocentricism of Disciplines and the Fish-Scale Model of Omniscience. Pages 328–348 in *Interdisciplinary Relationships in the Social Sciences*, edited by M. Sherif and C. W. Sherif. Chicago: Aldine Publishing Company.

Campbell, Eric G., Brian R. Clarridge, Manjusha Gokhale, Lauren Birenbaum, Stephen Hilgartner, Neil A. Holtzman, and David Blumenthal. 2002. Data Withholding in Academic Genetics: Evidence From a National Survey. *Journal of the American Medical Association* 287(4):473–480.

Campbell, R. 1989. *Biological Control of Microbial Plant Pathogens*. New York: Cambridge University Press.

Capshaw, James H., and Karen A. Rader. 1992. Big Science: Price to the Present. *Osiris*. 2nd Series 7:3–25.

Carley, H. Edwin. 1994. Curbing Crop Woes. *Farm Chemicals* (September): C18–C22

Charlesworth, Max, Lyndsay Farrall, Terry Stokes, and David Turnbull. 1989. *Life among the Scientists: An Anthropological Study of an Australian Scientific Community*. Melbourne, Australia: Oxford University Press.

Chefras, Jeremy. 1982. *Man-Made Life: An Overview of the Science, Technology, and Commerce of Genetic Engineering*. New York: Pantheon Books.

Chubin, Daryl E. and Edward J. Hackett. 1990. *Peerless Science: Peer Review and U.S. Science Policy*. Albany, N.Y.: State University of New York Press.

Cimons, Marlene and Paul Jacobs. 1999. Biotech Battlefield: Profits vs. Public. *Los Angeles Times* (February 21), A1, A16.

Clarke, Adele E. 1995. Research Materials and Reproductive Science in the United States, 1910–1940. Pages 119–135 in *Ecologies of Knowledge: Work and Politics in Science and Technology*, edited by Susan Leigh Star. Albany, N.Y.: SUNY Press.

———. 1991. Social Worlds/Arenas Theory as Organizational Theory. Pages 119–158 in *Social Organization and Social Process: Essays in Honor of Anselm L. Strauss*, edited by David Maines. Hawthorne, N.Y.: Aldine de Gruyter.

Clarke, Adele E., and Joan H. Fujimura, eds. 1992. *The Right Tools for the Job: At Work in Twentieth-Century Life Sciences*. Princeton, N.J.: Princeton University Press.

Cohen, Jon. 1997a. Please Pass the Data. *Science* 276 (June 27): 1961.

———. 1997b. PCR Patent Tangle Slows Quick Assay of HIV Levels. *Science* 276 (June 6): 1488–1491.

———. 1997c. Exclusive License Rankles Genome Researchers. *Science* 276 (June 6): 1489.

College of Agricultural and Life Science. 1997. The Institute for Pest and Pathogen Management. Brochure. University of Wisconsin-Madison (March).

Collins, H. M., and Trevor Pinch. 1993. *The Golem: What Everyone Should Know About Science*. Cambridge: Cambridge University Press.

Cook, R. James, and Kenneth F. Baker. 1983. *The Nature and Practice of Biological Control of Plant Pathogens*. St. Paul, Minn.: American Phytopathological Society.

Crease, Robert P. 1999. *Making Physics: A Biography of Brookhaven National Laboratory, 1946–1972*. Chicago: University of Chicago Press.

Curry, James, and Martin Kenney. 1990. Land Grant University-Industry Relations in Biotechnology: A Comparison with the Non-Land-Grant Research Universities. *Rural Sociology* 55(1): 44–57.

Denison, Niki. 1998. The Corporate Conundrum. *On Wisconsin* (Winter): 18–23, 53.

Ding, Dan. 1998. Rationality Reborn: Historical Roots of the Passive Voice in Scientific Discourse. Pages 117–135 in *Essays in the Study of Scientific Discourse: Methods, Practice and Pedagogy*, edited by John Battalio. Stamford, Conn.: Ablex Publishing Corp.

Doll, John H. 1998. The Patenting of DNA. *Science* 280 (May 1): 689–690.

Downey, Gary Lee. 1998. *The Machine in Me: An Anthropologist Sits Among Computer Engineers*. New York: Routledge.

———. 1995. The World of Industry-University-Government: Reimagining R&D as America. Pages 197–226 in *Technoscientific Imaginaries*, edited by George Marcus. Chicago: University of Chicago Press.

Downey, Gary Lee, and Joseph Dumit. 1997. *Cyborgs and Citadels:*

Anthropological Interventions in Emerging Sciences and Technologies. Santa Fe, N.M.: School of American Research Press.

Doyle, Jack. 1985. *Altered Harvest: Agriculture, Genetics and the Fate of the World's Food Supply*. New York: Viking.

Dubinskas, Frank A. 1988. Cultural Constructions: The Many Faces of Time. Pages 3–38 in *Making Time: Ethnographies of High-Technology Organizations*, edited by Frank A. Dubinskas. Philadelphia: Temple University Press.

Dunlap, Thomas R. 1981. *DDT: Scientists, Citizens, and Public Policy*. Princeton, N.J.: Princeton University Press.

Dworkin, Gerald. 1987. Commentary: Legal and Ethical Issues. *Science, Technology, and Human Values*, 12(1):63–64.

Eichenwald, Kurt. 1997. Push for Royalties Threatens Use of Down Syndrome Test. *New York Times*, May 23, A1, C3.

Eisenberg, Rebecca. 1989. Patents and the Progress of Science: Exclusive Rights and Experimental Use. *University of Chicago Law Review* 56: 1017.

Eliasoph, Nina. 1998. *Avoiding Politics: How Americans Produce Apathy in Everyday Life*. Cambridge: Cambridge University Press.

Enserink, Martin. 2000. Patent Office May Raise the Bar on Gene Claims. *Science* 287 (February 18): 1196–1197.

———. 1999. NIH Proposes Rules for Materials Exchange. *Science* 284 (May 28): 1445.

———. 1998a. Physicians Way of Scheme to Pool Icelanders' Genetic Data. *Science* 281 (August 14): 890–891.

———. 1998b. Opponents Criticize Iceland's Database. *Science* 282 (October 30): 859.

Epicentre Technologies. 1997. *Epicentre Catalog: Products for Molecular Biology*. Madison, Wisc.: Epicentre Technologies.

Epstein, Steven. 1996. *Impure Science: AIDS, Activism, and the Politics of Knowledge*. Berkeley: University of California Press.

Etzkowitz, Henry. 1997. The Entrepreneurial University and the Emergence of Democratic Corporatism. Pages 141–152 in *Universities and the Global Economy*, edited by Henry Etzkowitz and Loet Leydesdorff. London: Pinter.

———. 1995. Science as Intellectual Property. Pages 480–505 in *Handbook of Science and Technology Studies*, edited by Sheila Jasanoff, Gerald E. Markle, James Petersen, and Trevor Pinch. Thousand Oaks, Calif.: Sage.

———. 1994. Knowledge as Property: The Massachusetts Institute of Technology and the Debate over Academic Patent Policy. *Minerva* 32(4):383–421.

———. 1992. Redesigning "Solomon's House": The University and the

Internationalization of Science and Business. In *Denationalizing Science,* edited by Elisabeth Crawford et al. Dordrecht, Netherlands: Kluwer Academic Publishers.

———. 1989. Entrepreneurial Science in the Academy: A Case of the Transformation of Norms. *Social Problems* 36(1):14–29.

Etzkowitz, Henry, and Loet Leydesdorff. 1997. Introduction: Universities in the Global Economy. Pages 1–10 in *Universities and the Global Economy,* edited by Henry Etzkowitz and Loet Leydesdorff. London: Pinter.

Etzkowitz, Henry, and Lois S. Peters. 1991. Profiting from Knowledge: Organisational Innovations and the Evolution of Academic Norms. *Minerva* 29 (3): 133–166.

Etzkowitz, Henry, and Andrew Webster. 1998. Entrepreneurial Science: The Second Academic Revolution. Pages 21–46 in *Capitalizing Knowledge,* edited by Henry Etzkowitz, Andrew Webster, and Peter Healy. Albany, N.Y.: SUNY Press.

Etzkowitz, Henry, Andrew Webster, and Peter Healy. 1998. Introduction. Pages 1–20 in *Capitalizing Knowledge,* edited by Henry Etzkowitz, Andrew Webster, and Peter Healy. Albany, N.Y.: SUNY Press.

Food and Chemical News. 1996. Food Pathogen Detection Methods Improving, Research Report at FDA Forum. *Food and Chemical News* 43 (December 16): 38.

Fine, Gary Alan. 1996. *Kitchens: The Culture of Restaurant Work*. Berkeley: University of California Press.

Forsythe, Diana. 2001. *Studying Those Who Study Us: An Anthropologist in the World of Artificial Intelligence*. Stanford, Calif.: Stanford University Press.

Frickel, Scott. 1996. Engineering Heterogeneous Accounts: The Case of Submarine Thermal Reactor Mark-I. *Science, Technology, and Human Values* 21(1):28–53.

Friendly, Jock. 1996. New Anticoagulant Prompts Bad Blood Between Partners. *Science* 271 (March 29): 1800–1801.

Fuchs, Stephan. 1992. *The Professional Quest for Truth: A Social Theory of Science and Knowledge*. Albany, N.Y.: State University of New York Press.

Fujimura, Joan H. 1996. *Crafting Science: A Sociohistory of the Quest for the Genetics of Cancer*. Cambridge, Mass.: Harvard University Press.

———. 1992. Crafting Science: Standardized Packages, Boundary Objects, and "Translation." Pages 168–214 in *Science as Practice and Culture,* edited by Andrew Pickering. Chicago: University of Chicago Press.

———. 1991. On Methods, Ontologies, and Representation in the

Sociology of Science: Where Do We Stand? Pages 207–248 in *Social Organization and Social Process: Essays in Honor of Anselm L. Strauss,* edited by David Maines. Hawthorne, N.Y.: Aldine de Gruyter.

———. 1988. The Molecular Biological Bandwagon in Cancer Research: Where Social Worlds Meet. *Social Problems* 35(3):261–283.

———. 1987. Constructing "Do-able" Problems in Cancer Research: Articulating Alignment. *Social Studies of Science* 17:257–293.

Gallepp, George. 1994. The Hunt for the Wunderbug. *Science Report,* College of Agricultural and Life Sciences, University of Wisconsin–Madison, 25–31.

Gardner, William, and Joseph Rosenbaum. 1998. Database Protection and Access to Information *Science* 281 (August 7): 786–787.

Gaudilliere, Jean-Paul, and Iliana Lowy. 1998a. General Introduction. Pages 3–15 in *The Invisible Industrialist: Manufactures and the Construction of Scientific Knowledge,* edited by Jean-Paul Gaudilliere and Iliana Lowy. London: Macmillan.

———. 1998b. Disciplining Cancer: Mice and the Practice of Genetic Purity. Pages 209–249 in *The Invisible Industrialist: Manufactures and the Construction of Scientific Knowledge,* edited by Jean-Paul Gaudilliere and Ilana Lowy. London: Macmillan.

Gaudilliere, Jean-Paul, and Ilana Lowy, eds. 1998. *The Invisible Industrialist: Manufactures and the Construction of Scientific Knowledge.* London: Macmillan.

Geiger, Roger L. 1993. *Research and Relevant Knowledge: American Research Universities Since World War II.* New York: Oxford University Press.

———. 1992. Science, Universities, and National Defense, 1945–1970. *Osiris.* 2nd series 7:26–48.

———. 1986. *To Advance Knowledge: The Growth of American Research Universities.* New York: Oxford University Press.

Giddens, Anthony. 1984. *The Constitution of Society.* Berkeley: University of California Press.

Gieryn, Thomas F. 1998. Biotechnology's Private Parts (and Some Public Ones). Pages 219–253 in *Private Science: Biotechnology and the Rise of the Molecular Sciences,* edited by Arnold Thackray. Philadelphia: University of Pennsylvania Press.

Gilbert, Gregory S., Jo Handelsman, and Jennifer L. Parke. 1994. Root Camouflage and Disease Control. *Phytopathology* 84:222–225.

Gold, S. E., and H. J. Gold. 1983. Some Elements of a Model to Improve Productivity of Interdisciplinary Groups. Pages 86–101 in *Managing Interdisciplinary Research,* edited by S. R. Epton, R. L. Payne, and A. W. Pearson. New York: John Wiley and Sons.

Goodell, Rae. 1979. Public Involvement in the DNA Controversy: The

Case of Cambridge, Massachusetts. *Science, Technology, and Human Values* 27:36–43.

Goodman, Robert, Jo Handelsman, and Craig Grau. n.d. Development of Biocontrol for Root Diseases of Alfalfa for Wisconsin. Research proposal, Department of Plant Pathology, University of Wisconsin–Madison.

Gramsci, Antonio. 1971. *Selections from the Prison Notebooks of Antonio Gramsci*. Edited and translated by Quintin Hoare and Geoffrey Nowell Smith. New York: International Publishers.

Greathead, David J. 1994. History of Biological Control. *Antenna* 18(4): 187–199.

Grint, Keith, and Steve Woolgar. 1997. *The Machine at Work: Technology, Work, and Organization*. Cambridge, England: Polity Press.

Gross, Paul R., and Norman Levitt. 1994. *Higher Superstition: The Academic Left and its Quarrels with Science*. Baltimore: Johns Hopkins University Press.

Groves, Martha. 1998. Working out the Bugs. *Los Angeles Times*, May 4, D1, D7.

Hage, Jerald, and J. Rogers Hollingsworth. 2000. A Strategy for the Analysis of Idea Innovation Networks and Institutions. *Organization Studies* 21(5):971–1004.

Hagendijk, Rob. 1990. Structuration Theory, Constructivism, and Scientific Change. In *Theories of Science in Society*, edited by Susan E. Cozzens and Thomas F. Gieryn. Bloomington: Indiana University Press.

Hagman, Michael. 1999. EPA, Critics Soften Stance on Pesticidal Plants. *Science* 284 (April 9): 249.

Haiyin, He, Laura A. Silo-Suh, Jo Handelsman, and Jon Clardy. 1994. Zwittermicin A, an Antifungal and Plant Protection Agent from *Bacillus cereus*. *Tetrahedron Letters* 35:2499–2502.

Handelsman, Jo. 1994. A Regional Approach to Biocontrol for Alfalfa Root Diseases. North Central Sustainable Agriculture Preproposal.

Handelsman, Jo, Robert M. Goodman, and David W. Johnson. 1995. Biocontrol of Alfalfa Diseases with *Bacillus cereus*. Grant Proposal. Department of Plant Pathology, University of Wisconsin–Madison. April 15.

Handelsman, Jo, Ellen H. Mester, and Lynn Wunderlich. 1989. Biological Control of Damping Off and Root Rot and Inoculum Preparation Therefrom. *United States Patent* 4,877,738, October 31.

Handelsman, Jo, Sandra Raffel, Ellen H. Mester, Lynn Wunderlich, and Craig Grau. 1990. Biological Control of Damping-Off of Alfalfa Seedlings with *Bacillus cereus* UW85. *Applied and Environmental Microbiology* 56 (March): 713–718.

Handelsman, Jo, and Eric Stabb. 1996. Biocontrol of Soilborne Plant Pathogens. *The Plant Cell* 8:1855–1869.

Handelsman, Jo et al. 1994. Improving Alfalfa Health in Wisconsin: An Approach to Improving Alfalfa Health in Wisconsin Based on Exploiting Genetic Diversity and Empowering Local Business. Proposal for funding through the CALS Research Division Interdisciplinary Research Program.

Harrington, Jerry. 1996. The Midwest Agricultural Chemical Association: A Regional Study of an Industry on the Defensive. *Agricultural History* 70(2):415–438.

Harrison, Bennett. 1994. *Lean and Mean: The Changing Landscape of Corporate Power in the Age of Flexibility*. New York: Basic Books.

Hart, David M. 1998. *Forged Consensus: Science, Technology, and Economic Policy in the United States, 1921–1953*. Princeton, N.J.: Princeton University Press.

Hayley, Denis. 1994. Paris Green to Green Revolution. Farm Chemicals (September): C2–C4.

———. 1984. Focus on the Ag Chemical Industry. *Farm Chemicals* (December): 52–55.

Heffernan, William D. 2000. Pages 61–76 in *Hungry for Profit: The Agribusiness Threat to Farmers, Food, and the Environment*, edited by Fred Magdoff, John Bellamy Foster, and Frederick H. Buttel. New York: Monthly Review Press.

Heller, Michael A., and Rebecca S. Eisenberg. 1998. Response. *Science* 281 (July 24): 517.

Hendrickson, John A., G.W. Stevenson, Eric Stabb, and Jo Handelsman. n.d. Bioregional Biocontrol: A New Paradigm for Plant Health, Technology Transfer and Rural Community Development: Executive Summary. Unpublished.

Herrington, Tyanna K. 1999. Who Owns My Work? The State of Work for Hire for Academics in Technical Communication. *Journal of Business and Technical Communication* 13(2):125–153.

———. 1998. The Interdependency of Fair Use and the First Amendment. *Computers and Composition* 15:125–143.

Hess, David. 2000. Ethnography and the Development of Science and Technology Studies. Pages 234–245 in *Handbook of Ethnography*, edited by Paul Atkinson, Amanda Coffey, Sara Delamont, John Lofland, and Lyn Lofland. Thousand Oaks, Calif.: Sage Publications.

———. 1997a. *Science Studies: An Advanced Introduction*. New York: NYU Press.

———. 1997b. If You Are Thinking of Living in STS: A Guide for the Perplexed. Pages 143–164 in *Cyborgs and Citadels: Anthropological*

Interventions in Emerging Sciences and Technologies, edited by Gary Downey and Joseph Dumit. Santa Fe, N.M.: School of American Research Press.

———. 1995. *Science and Technology in a Multicultural World*. New York: Columbia University Press.

———. 1992. Introduction: The New Ethnography and the Anthropology of Science and Technology. *Knowledge and Society* 9:1–26.

Hilgartner, Stephen. 1998. Data Access Policy in Genome Research. Pages 202–218 in *Private Science: Biotechnology and the Rise of the Molecular Sciences,* edited by Arnold Thackray. Philadelphia: University of Pennsylvania Press.

Hilgartner, Stephen, and Sherry I. Brandt-Rauf. 1994. Data Access, Ownership, and Control: Toward Empirical Studies of Access Practices. *Knowledge* 15(4):355–372.

Hughes, Thomas P. 1987. The Evolution of Large Technological Systems. Pages 51–82 in *The Social Construction of Technological Systems,* edited by Wiebe E. Bijker, Thomas P. Hughes, and Trevor Pinch. Cambridge, Mass.: MIT Press.

Invitrogen. 1997. *Invitrogen: Product Catalog*. Carlsbad, Calif.: Invitrogen.

Jacobs, Paul. 1999. Stakes Are Huge in Genentech Patent Trial over Growth Drug. *Los Angeles Times,* May 3, C1, C8.

Jacobs, Paul, and Peter G. Gosselin. 2000. Experts Fret Over Effect of Gene Patents on Research. *Los Angeles Times,* February 28, pp. A1, A16.

Jasanoff, Sheila, Gerald E. Markle, James C. Petersen, and Trevor Pinch, eds. 1995. *Handbook of Science and Technology Studies*. Thousand Oaks, Calif.: Sage Publications, Inc.

Judson, Horace Freeland. 1992. A History of the Science and Technology Behind Gene Mapping and Sequencing. Pages 36–79 in *The Code of Codes: Scientific and Social Issues in the Human Genome Project,* edited by Daniel J. Kevles and Leroy Hood. Cambridge, Mass.: Harvard University Press.

Kaiser, Jocelyn. 1997. Academia Wins a Round on Raw Data. *Science* 277 (August 8): 758.

Kenney, Martin. 1985. *Biotechnology: The University-Industrial Complex*. New Haven, Conn.: Yale University Press.

Keller, Evelyn Fox. 1990. Physics and the Emergence of Molecular Biology: A History of Cognitive and Political Synergy. *Journal of the History of Biology* 23(3):389–409.

Kevles, Daniel J., and Gerald L. Geison. 1995. The Experimental Life Sciences in the Twentieth Century. *Osiris* 2nd Series (10):97–121.

Klein, Hans, and Daniel Lee Kleinman. 2002. The Social Construction of Technology: Structural Considerations. *Science, Technology, and Human Values* 27(1):28–52.

Kleinman, Daniel Lee. 1998a. Untangling Context: Understanding a University Laboratory in the Commercial World. *Science, Technology, and Human Values* 23(3):285–314.

———. 1998b. Pervasive Influence: Intellectual Property, Industrial History, and University Science. *Science and Public Policy* 25(2): 95–102.

———. 1995. *Politics on the Endless Frontier: Postwar Research Policy in the United States*. Durham, N.C.: Duke University Press.

Kleinman, Daniel Lee, and Jack Kloppenburg. 1988. Biotechnology and University-Industry Relations: Policy Issues in Research and Ownership of Intellectual Property at a Land Grant University. *Policy Studies Journal* 17(1):83–96.

Kleinman, Daniel Lee, and Steven P. Vallas. 2001. Science, Capitalism, and the Rise of the "Knowledge Worker": The Changing Structure of Knowledge Production in the United States. *Theory and Society* 30: 451–492.

Kloppenburg, Jack. 2000. Biopiracy, Witchery, and the Fables of Ecoliberalism. *Peace Review* 12(4):509–516.

———. 1990. No Hunting: Scientific Poaching and Global Biodiversity. *Z Magazine* (September): 104–108.

Kloppenburg, Jack R., Jr. 1988. *First the Seed: The Political Economy of Plant Biotechnology, 1492–2000*. New York: Cambridge University Press.

Kloppenburg, Jack R., and Daniel Lee. Kleinman. 1987. Seed Wars: Common Heritage, Private Property, and Political Strategy. *Socialist Review* 95:7–41

Knorr Cetina, Karin. 1999. *Epistemic Cultures: How the Sciences Make Knowledge.* Cambridge, Mass.: Harvard University Press.

———. 1981. *The Manufacture of Knowledge: An Essay in the Constructivist and Contextual Nature of Science*. Oxford: Pergamon.

———. 1995. Laboratory Studies: The Cultural Approach to the Study of Science. Pages 140–166 in *Handbook of Science and Technology Studies*, edited by Sheila Jasanoff, Gerald E. Markle, James C. Peterson, and Trevor Pinch. Thousand Oaks, Calif.: Sage.

———. 1985. Germ Warfare. *Social Studies of Science* 15:577–585.

———. 1982. Scientific Communities or Transepistemic Arenas of Research? A Critique of Quasi-Economic Models of Science. *Social Studies of Science* 12:101–130.

Kohler, Robert E. 1994. *Lord of the Fly: Drosophila Genetics and the Experimental Life*. Chicago: University of Chicago Press.

———. 1991. Systems of Production: Drosophila, Neurospora, and Biochemical Genetics. *HSPS* 22(1):87–130.

———. 1990. *Partners in Science: Foundations and Natural Scientists, 1900–1940*. Chicago: University of Chicago Press.

Kohn, Gustave. 1987. Agriculture, Pesticides, and the American Chemical Industry. In *Silent Spring Revisited,* edited by Gino Marco, Robert Hollingsworth, and William Durham. Washington, D.C.: American Chemical Society.

Krimsky, Sheldon, J. Ennis, and R. Weissman. 1991. Academic-Corporate Ties in Biotechnology: A Quantitative Study. *Science, Technology, and Human Values* 16:275–288.

Krohn, Wolfgang and Wolf Schafer. 1983. Agricultural Chemistry: The Origin and Structure of a Finalized Science. Pages 17–52 in *Finalization in Science: The Social Organization of Scientific Progress,* edited by Wolf Schafer. Boston: D. Reidel Publishing Company.

Lagemann, Ellen Condliffe. 1989. *The Politics of Knowledge: The Carnegie Corporation, Philanthropy, and Public Policy*. Chicago: University of Chicago Press.

Lambrecht, Bill. 1999. Judge Assails Secrecy, Halts Yellowstone "Bio-prospecting"; Products are Worth Millions. *St. Louis Post-Dispatch,* March 26, A1.

Latour, Bruno. 1988. *The Pasteurization of France*. Cambridge, Mass.: Harvard University Press.

———. 1987. *Science in Action*. Cambridge, Mass.: Harvard University Press.

———. 1983. Give Me a Laboratory and I Will Raise the World. Pages 141–170 in *Science Observed: Perspectives on the Social Study of Science,* edited by Karin Knorr and Michael Mulkay. Los Angeles: Sage.

Latour, Bruno and Steve Woolgar. 1986 [1979]. *Laboratory Life: The Construction of Scientific Facts*. Princeton, N.J.: Princeton University Press.

Law, John. 1991a. Introduction: Monsters, Machines, and Sociotechnical Relations. Pages 1–23 in *A Sociology of Monsters: Essays on Power, Technology, and Domination,* edited by John Law. London: Routledge.

———. 1991b. Power, Discretion, and Strategy. Pages 165–191 in *A Sociology of Monsters: Essays on Power, Technology, and Domination,* edited by John Law. London: Routledge.

Law, John, and Michel Callon. 1992. The Life and Death of an Aircraft: A Network Analysis of Technological Change. Pages 21–52 in *Shaping Technology, Building Society,* edited by Wiebe Bijker and John Law. Cambridge, Mass.: MIT Press.

Law, John, and John Hassard. 1999. *Actor Network Theory and After*. London: Blackwell Publishers.

Leslie, Stuart W. 1994. Science and Politics in Cold War America. Pages 199–233 in *The Politics of Western Science, 1640–1990,* edited by Margaret C. Jacobs. Atlantic Highlands, N.J.: Humanities Press.

———. 1993. *The Cold War and American Science: The Military-Industrial-Academic Complex at MIT and Stanford*. New York: Columbia University Press.

Lewis, Karen Seashore, and Melissa S. Anderson. 1998. The Changing Context of Science and University-Industry Relations. Pages 73–91 in *Capitalizing Knowledge,* edited by Henry Etzkowitz, Andrew Webster, and Peter Healy. Albany, N.Y.: SUNY Press.

Lindberg, Leon. 1982. The Problems of Economic Theory in Explaining Economic Performance. *Annals of the American Academic of Political and Social Sciences* 459:14–27.

Lowen, Rebecca S. 1997. *Creating the Cold War University: The Transformation of Stanford*. Berkeley: University of California Press.

Luesby, Jenny. 1996. Quest for Faster Growth in a Stagnating Market—The Ciba-Sandoz Alliance in Agrochemicals Reflects an Urgent Need for Consolidation. *Financial Times,* March 19, 25.

Lukes, Steven. 1974. *Power a Radical View*. London: Macmillan.

Lynch, Michael. 1993. *Scientific Practice and Ordinary Action: Ethnomethodology and Social Studies of Science*. New York: Cambridge University Press.

———. 1985. *Art and Artifact in Laboratory Science: A Study of Shop Work and Shop Talk in a Research Laboratory*. London: Routledge.

MacKenzie, Donald, and Judy Wajcman. 1985. Introductory Essay: The Social Shaping of Technology. Pages 2–25 in *The Social Shaping of Technology,* edited by Donald MacKenzie and Judy Wajcman. Buckingham, England: Open University Press.

MacKenzie, Michael, Peter Keating, and Alberto Cambrosio. 1990. Patents and Free Scientific Information in Biotechnology: Making Monoclonal Antibodies Proprietary. *Science, Technology, and Human Values* 15:65–83.

Malakoff, David. 1999a. Plan to Import Exotic Beetle Drives Some Scientists Wild. *Science* 284 (May 21): 1255.

———. 1999b. Fighting Fire with Fire. *Science* 285 (September 17): 1841–1843.

Marcus, George E. 1995. Ethnography in/of the World System: The Emergence of Multi-Sited Ethnography. *Annual Review of Anthropology* 24: 95–117.

Marshall, Eliot. 2000. Patent on HIV Receptor Provokes an Outcry. *Science* 287 (February 25): 1375, 1377.

———. 1999. Two Former Grad Students Sue over Alleged Misuse of Ideas. *Science* 284 (April 23): 562, 563.

———. 1997a. Secretness Found Widespread in Life Sciences. *Science* 276 (April 25): 525.

———. 1997b. The Mouse that Prompted a Roar. *Science* 277 (July 4): 24–25.
———. 1997c. Battling Over Basics. *Science* 277 (July 4): 25.
———. 1997d. Whose DNA is it, Anyway? *Science* 278 (October 24): 564–567.
———. 1997e. Gene Prospecting in Remote Populations. *Science* 278 (October 24): 565.
———. 1997f. Tapping Iceland's DNA. *Science* 278 (October 24): 566.
———. 1997g. Need a Reagent? Just Sign Here. . . . *Science* 278 (October 10): 212–213.
———. 1997h. Snipping Away at Genome Patenting. *Science* 277 (September 19): 1752–1753.
———. 1997j. Is Data-Hoarding Slowing the Assault on Pathogens. *Science* 275 (February 7): 777–780.
———. 1997k. Companies Rush to Patent DNA. *Science* 275 (February 7): 780–781.
———. 1997l. Sequencers Call for Faster Data Release. *Science* 276 (May 23): 1189–1190.
———. 1997m. "Playing Chicken" over Gene Markers. *Science* 278 (December 19): 2046–2048.
———. 1995. Suit Alleges Misuse of Peer Review. *Science* 270 (December 22): 1912–1914.
———. 1984. The Enduring Problem of Pesticide Misuse. *Technology Review* 87 (February): 10.
Marshall, Eliot, and Pallava Bagla. 1997. India Applauds Patent Reversal. *Science* 277 (September 5): 1429.
Martin, Brian. 1999. Suppression of Dissent in Science. *Social Problems and Public Policy* 7:105–135.
———. 1995. Against Intellectual Property. *Philosophy and Social Action* 21(3):7–22.
———. 1988. Analyzing the Fluoridation Controversy: Resources and Structures. *Social Studies of Science* 18:331–63.
Martin, Emily. 1994. *Flexible Bodies: The Role of Immunity in American Culture from the Days of Polio to the Age of AIDS*. Boston: Beacon Press.
Mauer, Stephen M., and Suanne Scotchmer. 1999. Database Protection: It Is Broken and Should We Fix It. *Science* 284 (May 14): 1129, 1130.
McMath, Robert C., Jr., Ronald H. Bayor, James E. Brittain, Lawrence Foster, August Giebelhaus, and Germaine M. Reed. 1985. *Engineering the New South: Georgia Tech, 1885–1985*. Athens: University of Georgia Press.
McNew, George L. 1959. Landmarks during a Century of Professional Use of Chemicals to Control Plant Diseases. Pages 42–54 in *Plant*

Pathology: Problems and Progress, 1908–1958, edited by C. S. Holton, G. W. Fischer, R. W. Fulton, Helen Hart, and S. E. A. McCallan. Madison: University of Wisconsin Press.

Merton, Robert K. 1973 [1942]. The Normative Structure of Science. Pages 254–278 in *The Sociology of Science: Theoretical and Empirical Investigations*. Chicago: University of Chicago Press.

Microbial ID, Inc. www.dca.net/midi/ Taken from website, October 9, 1997.

Mills, C. Wright. 1959. *The Sociological Imagination*. New York: Oxford University Press.

Milner, Jocelyn L., Laura Silo-Suh, Julie C. Lee, Haiyin He, Jon Clardy, and Jo Handelsman. 1996. Production of Kanosamine by *Bacillus cereus* UW85. *Applied and Environmental Microbiology* 62(8):3061–3065.

Milner, Jocelyn, Elizabeth Stohl, and Jo Handelsman. 1996. Zwittermicin A Resistance Gene from *Bacillus cereus*. *Journal of Bacteriology* 178 (14):4266–4272.

Mishkin, Barbara. 1995. Urgently Needed: Policies on Access to Data by Erstwhile Collaborators. *Science* 270 (November 10): 927–928.

Moffat, Anne Simon. 1991. Research on Biological Control Pest Control Moves Ahead. *Science* 252 (April 12): 211–212.

Moore, Kelly. 1996. Organizing Integrity: American Science and the Creation of Public Interest Organizations. *American Journal of Sociology* 101:1592–1627.

______. 1995. When Inside and Outside Matter: Scientists, Activism, and Changes in Scientific Knowledge. Paper presented at the Annual Meeting of the Society for the Social Studies of Science, October.

Morrison, Jessica. 1989. Biological Control Turns 100 This Year. *Agricultural Research* (March): 4–8.

Mueller, Janice M. 2001. No "Dilettante Affair": Rethinking the Experimental Use Exception to Patent Infringement for Biomedical Research Tools. *Washington Law Review* 76:1.

Myers, Greg. 1995. From Discovery to Invention: The Writing and Rewriting of Two Patents. *Social Studies of Science* 25:57–105.

National Science Board. 1993. *Science and Engineering Indicators—1993*. Washington, D.C.: Government Printing Office.

Nature. 1995. Roche Asked for More Taq Patent Evidence. *Nature* 374 (March 9): 108.

Nelkin, Dorothy, and Richard Nelson. 1987. Commentary: University-Industry Alliances. *Science, Technology, and Human Values* 12(1): 65–74.

Nelson, Lita. 1998. The Rise of Intellectual Property Protection in the American University. *Science* 279 (March 6): 1460, 1461.

Noble, David. 1984. *Forces of Production: A Social History of Industrial Automation*. New York: Knopf.

———. 1983. Academia Incorporated. *Science for the People* (January/February): 7–11, 50–52.

———. 1977. *America by Design: Science, Technology, and the Rise of Corporate Capitalism*. New York: Oxford University Press.

Ordish, George. 1976. *The Constant Pest: A Short History of Pests and their Control*. New York: Charles Scribner's Sons.

Osburn, Robert M., Jocelyn Milner, Edward S. Oplinger, R. Stewart Smith, and Jo Handelsman. 1995. Effect of *Bacillus cereus* UW85 on the Yield of Soybean at Two Field Sites in Wisconsin. *Plant Disease* 79: 551–556.

Osteen, Craig D., and Philip I. Szmedra. 1989. *Agricultural Pesticide Use Trends and Policy Issues* (Agricultural Economic Report Number 622). Washington, D.C.: United States Department of Agriculture, Economic Research Service.

Oudshoorn, Nelly. 1990. On the Making of Sex Hormones: Research Materials and the Production of Knowledge. *Social Studies of Science* 20:5–33.

Owen-Smith, Jason. 2001. Managing Laboratory Work through Skepticism: Processes of Evaluation and Control. *American Sociological Review* 66 (June): 427–452.

PR Newswire. 1993. Promega Corp. Challenges Key Patent Held by Hoffmann-La Roche. *PR Newswire*, April 15. Taken from Nexus.

Packer, Kathryn, and Andrew Webster. 1996. Patenting Culture in Science: Reinventing the Scientific Wheel of Credibility. *Science, Technology, and Human Values* 21(4):427–453.

———. 1995. Inventing Boundaries: The Prior Art of the Social World. *Social Studies of Science* 25:107–117.

Palladino, Paulo. 1996. *Entomology, Ecology and Agriculture: The Making of Scientific Careers in North America*. Amsterdam: Harwood Academic Publishers.

Parker, David L. 1994. Patent Infringement Exemptions for Life Science Research. *Houston Journal of International Law* 16: 615.

Pennisi, Elizabeth. 1998. Lawsuit Targets Yellowstone Bug Deal. *Science* 279 (March 13): 1624.

———. 1997. Merck Gives Researchers Knockout Deal. *Science* 276 (April 25): 527.

Perkins, John H. 1982. *Insects, Experts, and the Insecticide Crisis: The Quest for New Pest Management Strategies*. New York: Plenium.

Peters, Lois S., and Henry Etzkowitz. 1991. University-Industry Connections and Academic Values. *Technology in Society* 12:427–440.

Pfeffer, Jeffrey, and Gerald Salancik. 1978. *The External Control of Organizations: A Resource Dependence Perspective*. New York: Harper and Row.

Phipps, P. M. 1992. Peanut (*Arachis hypogaea "NC9")* Sclerotinia blight; *Sclerotinia minor*. *Biological and Cultural Tests for Control of Plant Diseases* 7:60.

Pickering, Andrew. 1995. *The Mangle of Practice: Time, Agency, and Science*. Chicago: University of Chicago Press.

Pickering, Andrew, ed. 1992. *Science as Practice and Culture*. Chicago: University of Chicago Press.

Pollack, Andrew. 2000. Bioprospecting Deal Gains in Yellowstone. *New York Times* April 25, F5.

———. 1999. Biological Products Raise Genetic Ownership Issues. *New York Times*, A1, C4.

Powell, Walter W. 1998. Learning from Collaboration: Knowledge and Networks in the Biotechnology and Pharmaceutical Industrials. *California Management Review* 40:3.

Rabinow, Paul. 1996. *Making PCR: A Story of Biotechnology*. Chicago: University of Chicago Press.

Rabkin, Yakov M. 1992. Rediscovering the Instrument: Research, Industry, and Education. Pages 57–72 in *Invisible Connections: Instruments, Institutions, and Science*, edited by Robert Bud and Susan Cozzens. Bellingham, Wash.: International Society for Optical Engineering.

Raffel, Sandra J., Eric V. Stabb, Jocelyn Milner, and Jo Handelsman. 1996. Genotype and Phenotypic Analysis of Zwittermicin A-Producing Strains of *Bacillus cereus*. *Microbiology* 142:3425–3436.

Rappert, Brian, and Andrew Webster. 1997. Regimes of Ordering: The Commercialization of Intellectual Property in Industrial-Academic Collaborations. *Technology Analysis and Strategic Management* 9(2): 115–130.

Restivo, Sal. 1995. The Theory Landscape in Science Studies. Pages 95–110 in *Handbook of Science and Technology Studies*, edited by Sheila Jasanoff, Gerald E. Markle, James C. Petersen, and Trevor Pinch. Thousand Oaks, Calif.: Sage.

Reuters. 2001. US Issues Stiffer Regulations on Frivolous Patenting of Genes. *New York Times* January 6, C3.

———. 1993. Promega Contests Hoffmann-La Roche Patent. *Reuters*, April 15. Taken from Nexus.

Rhodes, Gary, and Sheila Slaughter. 1997. Academic Capitalism, Managed Professionals, and Supply-Side Higher Education. *Social Text* 15(2):9–38.

Rosenberg, Charles. 1979. Toward an Ecology of Knowledge: On

Discipline, Context, and History. Pages 440–455 in *The Organization of Knowledge in Modern America: 1860–1920,* edited by Alexandra Oleson and John Voss. Baltimore: Johns Hopkins University Press.

Rossiter, Margaret W. 1979. The Organization of the Agricultural Sciences. Pages 211–248 in *The Organization of Knowledge in Modern America: 1860–1920,* edited by Alexandra Oleson and John Voss. Baltimore: Johns Hopkins University Press.

Russell, M.G. 1983. Peer Review in Interdisciplinary Research: Flexibility and Responsiveness. In *Managing Interdisciplinary Research,* edited by S.R. Epton, R.L. Payne, and A.W. Pearson. New York: John Wiley and Sons.

Samuelson, Pamela. 1987. Innovation and Competition: Conflicts over Intellectual Property in New Technologies. *Science, Technology, and Human Values* 11(1):6–21.

Sawyer, Richard C. 1996. *To Make a Spotless Orange: Biological Control in California.* Ames, Iowa: Iowa State University Press.

Saxberg, Borje O. and William T. Newell. 1983. Interdisciplinary Research in the University: Need for Managerial Leadership. In *Managing Interdisciplinary Research,* edited by S. R. Epton, R. L. Payne, and A. W. Pearson. New York: John Wiley and Sons.

Saxegaard, F., I. Baess, and E. Jantzen. 1988. Characterization of Clinical Isolates of Mycobacterium paratuberculosis by DNA–DNA Hybridization and Cellular Fatty Acid Analysis. *Acta Pathologica Microbiologica et Immunologica Scadinavica* (APMIS) 96(6):497–502.

Saxenian, Annalee. 1994. *Regional Advantage: Culture and Competition in Silicon Valley and Route 128.* Cambridge, Mass.: Harvard University Press.

Scher, F. M., and J. R. Castagno. 1986. Biocontrol: A View from Industry. *Canadian Journal of Plant Pathology* 8:222–224.

Science. 1997a. Gene Fragments Patentable, Official Says. *Science* 275 (February 21): 1055.

———. 1997b. Renewed Fight over Gene Patent Policy. *Science* 276 (April 11): 187.

———. 1997c. Academy Joins Debate over DNA Patents. *Science* 277 (July 4): 41.

———. 1997d. Yellowstone Opens the Gates to Biotech. *Science* 277 (August 22): 1027.

———. 1996. Company Secrets Don't Stop Science. *Science* 271 (March 8): 1367.

Scotsman, The. 1997. Where Should the Line be Drawn on Who Owns Life Itself? *The Scotsman,* July 23, 1997.

Scott, Pam. 1991. Levers and Counterweights: A Laboratory that Failed to Raise the World. *Social Studies of Science* 21: 7–35.

Serageldin, Ismail. 1999. Biotechnology and Food Security in the 21st Century. *Science* 285 (July 16): 387–389.

Service, Robert F. 1999. Taq Polymerase Patent Ruled Invalid. *Science* 286 (December 17): 2251, 2253.

———. 1998. Seed-Sterilizing "Terminator Technology" Sows Discord. *Science* 282 (October 30): 850, 851.

Servos, John W. 1996. Engineers, Businessmen, and the Academy: The Beginnings of Sponsored Research at the University of Michigan. *Technology and Culture* 37(4):721–762.

———. 1994. Changing Partners: The Mellon Institute, Private Industry, and the Federal Patron. *Technology and Culture* 35(2):221–257.

Shapin, Steven. 1994. *A Social History of Truth: Civility and Science in Seventeenth Century England*. Chicago: University of Chicago Press.

———. 1988. Following Scientists Around. *Social Studies of Science* 18: 533–550.

Shinn, Terry. 1997. Crossing Boundaries: The Emergence of Research-Technology Communities. Pages 85–96 in *Universities and the Global Economy,* edited by Henry Etzkowitz and Loet Leydesdorff. London: Pinter.

Shiva, Vanda. 1997. *The Plunder of Nature and Knowledge*. Boston: South End Press.

Shon, Melissa. 1994. The Brighter Spots. *(New York) Chemical Marketing Reporter*. 17 (April 25): 245. Taken from ABI Inform.

Shulman, Seth. 1999a. We Need New Ways to Own and Share Knowledge. *Chronicle of Higher Education* (19 February): A64.

———. 1999b. *Owning the Future: Inside the Battles to Control the New Assets—Genes, Software, Databases, and Technological Know-How—That Make Up the Lifeblood of the New Economy*. New York: Houghton Mifflin.

———. 1998. Cashing in on Medical Knowledge. *Technology Review* (March/April): 38–43.

Silo-Suh, Laura A., Benjamin J. Letherbridge, Sandra J. Raffel, Haiyin He, Jon Clardy, and Jo Handelsman. 1994. Biological Activities of Two Fungistatic Antibiotics Produced by *Bacillus cereus* UW85. *Applied and Environmental Microbiology* 60 (June): 2023–2030.

Sismondo, Sergio. 1996. *Science Without Myth: On Constructions, Reality, and Social Knowledge*. Albany, N.Y.: SUNY Press.

Slaughter, Sheila, and Larry L. Leslie. 1997. *Academic Capitalism: Politics, Policies, and the Entrepreneurial University*. Baltimore: Johns Hopkins University Press.

Slaughter, Sheila, and Gary Rhodes. 1996. The Emergence of a Competitiveness Research and Development Policy Coalition and the Commercialization of Academic Science and Technology. *Science, Technology, and Human Values* 21(3):303–339.

———. 1990. Renorming the Social Relations of Academic Science: Technology Transfer. *Educational Policy*. 4: 341–361.

Smith, John K. 1988. World War II and the Transformation of the American Chemical Industry. In *Science, Technology, and the Military,* edited by E. Mendelsohn, M. R. Smith, and P. Weingart. Dordrecht, Netherlands: Kluwer Academic Publishers

Smith, A. Ray, and Joel P. Seigel. 1996. Cellular Fatty Acid Analysis for the Classification and Identification of Bacteria. Pages 180–222 in *Automated Microbial Identification and Quantitation: Technologies for the 2000s,* edited by Wayne P. Olson. Buffalo Grove, Ill.: Interpharm Press, Inc.

Stabb, Eric V., Lynn M. Jacobson, and Jo Handelsman. 1994. Zwittermicin A–Producing Strains of *Bacillus cereus* from Diverse Soils. *Applied and Environmental Microbiology* 60(12):4404–4412.

Standard and Poor's. 1997. Industry Profile: Specialty Chemicals. *Industry Surveys*. March 20. Volume 1. New York.

Star, Susan Leigh, and James Griesemer. 1989. Institutional Ecology, "Translations" and Boundary Objects: Amateurs and Professionals in Berkeley's Museum of Vertebrate Zoology, 1907–1939. *Social Studies of Science* 19:387–420.

Stevenson, G. W., John Hendrickson, Jo Handelsman, and Eric Stabb. n.d. Bioregional Biocontrol of Alfalfa Disease: New Paradigms for Crop Health and Technology Transfer. Unpublished Status Report.

Stevenson, John H. 1959. The Beginnings of Plant Pathology in North America. Pages 14–23 in *Plant Pathology: Problems and Progress, 1908–1958,* edited by C. S. Holton, G. W. Fischer, R. W. Fulton, Helen Hart, and S. E. A. McCallan. Madison: University of Wisconsin Press.

Stohl, Elizabeth, Eric V. Stabb, and Jo Handelsman. 1996. Zwittermicin A and Biological Control of Oomycete Pathogens. Pages 475–479 in *Biology of Plant-Microbe Interactions: Proceedings of the 8th International Symposium on Molecular Plant-Microbe Interactions,* edited by Gray Stacey, Beth Mullin, and Peter M. Gresshoff. St. Paul, Minn.: International Society for Molecular Plant-Microbe Interactions.

Sutz, Judith. 1997. The New Role of the University in the Productive Sector. Pages 11–20 in *Universities and the Global Economy,* edited by Henry Etzkowitz and Loet Leydesdorff. London: Pinter.

Telitelman, Robert. 1989. *Gene Dreams: Wall Street, Academia, and the Rise of Biotechnology*. New York: Basic Books.

Todes, Daniel P. 1997. Pavlov's Physiology Factory. *Isis* 88:205–246.

Transpacific Media, Inc. 1997. Dr. DNA. May, Section 69, p. 56. Taken from Nexus.

Trapero-Caseas, Antonio, Walter J. Kaiser, and David M. Ingram. 1990. Control of Phythium Seed Rot and Preemergence of Damping-off of Chickpea in the U.S. Pacific Northwest and Spain. *Plant Disease* 74(8): 563–569.

Traweek, Sharon. 1988. *Beamtimes and Lifetimes: The World of High Energy Physics*. Cambridge, Mass.: Harvard University Press.

United Press International (UPI). 1997. BC Cycle. June 15. Taken from Nexus.

———. 1989. Resistant Varieties, Fungicides Can Ward Off Root Rot. August 21. Taken from Nexus.

———. 1987. Researchers recommend using treated alfalfa Seed. March 30. Taken from Nexus.

United States Patent and Trademark Office. 2001. Utility Examination Guidelines. *Federal Register* 66 (4) (January 5): 1092–1099.

Vaidhyanonthan, Siva. 2001. *Copyrights and Copywrongs: The Rise of Intellectual Property and How it Threatens Creativity*. New York: New York University Press.

Vasanen, O., and M. Salkinoja-Salonen. 1989. Use of Phage Typing and Fatty Acid Analysis for the Identification of Bacilli Isolated from Food Packaging Paper and Board Machines. *Systematic and Applied Microbiology* 12(1):103–111.

Vidhyasekaran, P. and M. Muthamilan. 1995. Developing Formulations of *Pseudomonas fluorescens* for Control of Chickpea Wilt. *Plant Disease* 79(8):782–786.

Vogel, Gretchen. 2000. Wisconsin to Distribute Embryonic Cell Lines. *Science* 287 (February 11): 948, 949.

Ward, Steven C. 1996. *Reconfiguring Truth: Postmodernism, Science Studies, and the Search for a New Model of Knowledge*. New York: Rowman and Littlefield Publishers, Inc.

Webster, Andrew. 1998. Strategic Alliances: Testing the Collaborative Limits. Pages 95–110 in *Capitalizing Knowledge,* edited by Henry Etzkowitz, Andrew Webster, and Peter Healy. Albany, N.Y.: SUNY Press.

———. 1994a. International Evaluation of Academic-Industry Relations: Contexts and Analysis. *Science and Public Policy* 21:72–78.

———. 1994b. University-Corporate Ties and the Construction of Research Agendas. *Sociology* 28:123–142.

———. 1990. The Incorporation of Biotechnology into Plant-Breeding in Cambridge. Pages 177–201 in *Deciphering Science and Technology: The Social Relations of Expertise,* edited by Ian Varcoe, Maureen McNeil, and Steven Yearly. London: Macmillan.

Webster, Andrew, and Henry Etzkowitz. 1998. Toward a Theoretical Analysis of Academic-Industrial Collaboration. Pages 47–72 in *Capitalizing Knowledge,* edited by Henry Etzkowitz, Andrew Webster, and Peter Healy. Albany, N.Y.: SUNY Press.

Webster, Andrew, and Kathryn Packer. 1997. When Worlds Collide: Patents in Public-Sector Research. Pages 47–59 in *Universities and the Global Economy,* edited by Henry Etzkowitz and Loet Leydesdorff. London: Pinter.

———. 1996. Patents and Technology Transfer in Public Sector Research: The Tension Between Policy and Practice. Pages 43–64 in *Barriers to International Technology Transfer,* edited by J. Kirkland. Dordrecht, Netherlands: Kluwer Academic Publishers.

Weil, Vivian. 1987. Introduction to a Special Section: Private Appropriation of Public Research. *Science, Technology, and Human Values* 12(1): 1–5.

Weiner, Charles. 1987. Patenting and Academic Research: Historical Case Studies. *Science, Technology, and Human Values* 12(1):50–62.

Wellman, R. H. 1959. Commercial Development of Fungicides. Pages 239–245 in *Plant Pathology: Problems and Progress, 1908–1958,* edited by C. S. Holton, G. W. Fischer, R.W. Fulton, Helen Hart, and S. E. A. McCallan. Madison: University of Wisconsin Press.

Whitley, Richard. 1984. *The Intellectual and Social Organization of the Sciences*. Oxford: Clarendon Press.

Williams, Paul H., and Melissa Marosy. 1986. *With One Foot in the Furrow: A History of the First Seventy-five Years of the Department of Plant Pathology at the University of Wisconsin–Madison*. Dubuque, Iowa: Kendall/Hunt Publishing Company.

Wolfsy, Leon. 1986. Biotechnology and the University. *Journal of Higher Education* 57(5):477–492.

Wright, Susan. 1994. *Molecular Politics: Developing American and British Regulatory Policy for Genetic Engineering, 1972–1982*. Chicago: University of Chicago Press.

———. 1986. Recombinant DNA Technology and Its Social Transformation, 1972–1982. *Osiris* 2nd Series 2:303–360.

Wynne, Brian. 1992. Representing Policy Constructions and Interests in SSK. *Social Studies of Science* 22:575–580.

Yoxen, Edward. 1983. *The Gene Business: Who Should Control Biotechnology*. New York: Oxford University Press.

Index